Electronics 2

Electronics 2

S. A. Knight
Lately Senior Lecturer in Mathematics & Electronic Engineering, Bedford College of Higher Education

Newnes
An imprint of Butterworth-Heinemann Ltd
Linacre House, Jordan Hill, Oxford OX2 8DP

A member of the Reed Elsevier group

OXFORD   LONDON   BOSTON
MUNICH   NEW DELHI   SINGAPORE   SYDNEY
TOKYO   TORONTO   WELLINGTON

First published as *Electronics for Technicians 2* 1978
Reprinted 1979
Second edition 1987
Reprinted 1989, 1990, 1991, 1993, 1994

© S. A. Knight 1979, 1987

All rights reserved. No part of this publication
may be reproduced in any material form (including
photocopying or storing in any medium by electronic
means and whether or not transiently or incidentally
to some other use of this publication) without the
written permission of the copyright holder except in
accordance with the provisions of the Copyright,
Designs and Patents Act 1988 or under the terms of a
licence issued by the Copyright Licensing Agency Ltd,
90 Tottenham Court Road, London, England W1P 9HE.
Applications for the copyright holder's written permission
to reproduce any part of this publication should be addressed
to the publishers

British Library Cataloguing in Publication Data
Knight, S. A.
   Electronics 2
   1. Electronic apparatus and appliances
   I. Title   II. Series
   621.381   TK7870   77-30752

ISBN 0 7506 0245 7

Printed and bound in Great Britain
by Hartnolls Ltd, Bodmin, Cornwall

Contents

PREFACE vii

1 THERMIONIC AND SEMICONDUCTOR THEORY
Structure of atoms 1
Conduction 3
Thermionic emission 3
Cathodes 4
Semiconductors 5
Impurity atoms 7
Problems for Section 1 8

2 SEMICONDUCTOR AND THERMIONIC DIODES
The junction diode 11
The thermionic diode 14
Comparisons 15
Problems for Section 2 16

3 APPLICATIONS OF SEMICONDUCTOR DIODES
Rectifier circuits 18
The half-wave rectifier 19
The full-wave rectifier 20
The bridge rectifier 21
Practical diode ratings 21
Smoothing circuits 23
The Zener diode 24
The Zener stabiliser 25
Zener protection 28
The varactor diode 29
Problems for Section 3 29

4 THE BIPOLAR TRANSISTOR
The junction transistor 31
Circuit configurations 33
Characteristic curves 35
Problems for Section 4 41

5 THE BIPOLAR TRANSISTOR AS AMPLIFIER
The general amplifier 43
The common-emitter amplifier 44
Using the characteristic curves 47
Leakage current 51
Setting the operating point 52
The transistor as a switch 55
Mutual conductance 55
Thermal runaway 56
Gain estimations 57
Problems for Section 5 58
Thermal runaway 55
Gain estimations 55
Problems for section 5 56

6 THE UNIPOLAR TRANSISTOR

The junction gate FET 60
The JUGFET as an amplifier 66
Load line analysis 71
The insulated gate FET 72
Problems for Section 6 74

7 STABILISED POWER SUPPLIES

Shunt regulation 78
Series regulation 79
Comparator methods 79
Protection 81
Problems for Section 7 81

8 LOGIC CIRCUITS

Positive and negative logic 84
Rules of circuit logic 84
Negation 87
Testing for logical equivalence 89
Logical relations and algebra 90
Problems for Section 8 94

9 COMBINATIONAL LOGIC GATES

Electronic gates 96
Electronic switching 97
Diode-resistor gates 98
Gate combinations 100
Diode-transistor logic 103
Transistor-transistor logic 103
Commercial integrated logic 105
Gate economy 110
Some hints about using TTL and CMOS 111
Problems for Section 9 112

10 KARNAUGH MAPPING

The map structure 114
Filling in the map 115
Truth table to K-map 116
Using the K-map 116
Rules for minimisation 118
Reading the K-map 119
Application to circuit design 120
Problems for Section 10 121

11 SEQUENTIAL LOGIC SYSTEMS

A transistor flip-flop 123
Triggering the flip-flop 125
Integrated systems 126
The clocked R–S flip-flop 127
The D-latch 128
The J-K flip-flop 128
Shift registers and counters 130
Problems for Section 11 139

SOLUTIONS TO PROBLEMS 140

APPENDIX 146

INDEX 147

Preface

This book has been revised and updated to cover the essential syllabus requirements of BTEC Unit U86/331 for Electronics 2. Although the guide syllabus has been followed to a large extent, there are occasionally some additional notes on particular topics where their inclusion seemed justified for the sake of clarity and completeness.

As an example of this, a completely separate chapter has been included on Karnaugh mapping, a subject not easily covered by a few passing references. Also, additions to the syllabus covering logic systems, both combinational and sequential, have been included with emphasis on the applications of the more popular commercial integrated packages from the 7400 TTL and 4000 CMOS families of logic devices. The unipolar (field effect) transistor is now treated at this level of the course in addition to the bipolar transistor, and a chapter on stabilised power supply systems has been added.

The book is best used in the order in which it is written. There are problems within the text of each chapter and at the end of each chapter. If these are worked methodically, the course will, I believe, proceed in a logical sequence, and at no point will a new or unfamiliar concept appear that has not already been explained, or is not in the process of being explained. Solutions are given to the problems and some of these contain additional information which can be treated as part of the main text.

A few errors which appeared in the earlier editions have now hopefully been eliminated. If any have slipped through I would be grateful to have them pointed out to me.

S. A. K.
Market Harborough

1 Thermionic and semiconductor theory

Aims: At the end of this Unit section you should be able to:
Understand the general structure of atoms.
Understand and describe the effects of thermionic emission.
Define the properties of conductors, insulators and semiconductors.
Name the majority charge carriers in p- *and* n-*type material.*
State the effect of temperature upon intrinsic conduction in semiconducting materials.

Electronics may be defined as *the study of the behaviour of electrons and the practical uses to which such study can be applied.* Up to about 25 years ago the term was associated almost exclusively with the theory and utilisation of thermionic valves, and all practical applications such as radio and television receivers, amplifiers, radar installations and control systems in industry depended upon the thermionic valve.

Over the intervening years the valve has been gradually replaced by the bipolar transistor, and this in turn is being replaced by field effect devices. Complicated circuits which, even with transistors, occupied a great amount of space are being replaced by integrated circuits which at the most occupy only a few cubic centimetres yet perform the work of thousands of discrete transistors and their associated circuit components.

Although thermionic valves are now obsolete in many applications, they remain superior in a number of high frequency and high power domains, and they are, of course, still found in considerable numbers in equipments such as older television receivers and industrial control systems. The cathode ray tube, the vital component of all television receivers and laboratory oscilloscopes, is a thermionic device. The principles involved are therefore worth some discussion, although the main emphasis will be upon semiconductors and associated devices. It is well to bear in mind that a grasp of the principles of thermionic valves is a very useful stepping stone to an understanding of semiconductors, particularly field effect devices where the analogous behaviour of valves and semiconductors is particularly well illustrated.

In this present Unit section we shall consequently introduce the basic theory not only of the semiconductor but also of the thermionic valve, to begin studying the operation and application of these electronic devices in general.

STRUCTURE OF ATOMS All matter exists in a solid, liquid or gaseous state. Matter in any of these states is made up of very small particles known as *molecules*, which in turn are composed of *atoms*. Atoms can be envisaged as minute planetary systems, having a nucleus or core which carries a positive charge of electricity, around which revolve small charges of negative electricity known as *electrons*. The mass of an electron is estimated to be about 9.1×10^{-31} kilograms (kg) and the charge it carries is about -1.6×10^{-19} coulomb (C). Both of these quantities are unimaginably small, but it is upon the electron, nevertheless, that the whole science of electronics depends. Normally an atom is electrically neutral, the effect of the negative charges carried by the revolving electrons being exactly balanced by the overall positive charge carried by the nucleus. The nucleus is not just a solid lump; it is made up of two other types of particles, *protons* and *neutrons*. Only the protons carry positive charges, the neutrons being without charge. The charge on each proton, since it neutralises the charge on an electron, is the same as that of an electron but of opposite sign. The proton, however, is about 1840 times as massive as an electron, so that a quick calculation gives its mass as 1.67×10^{-27} kg. Do the calculation for yourself.

The revolving electrons are pictured as moving in elliptical orbits around the nucleus, held in their respective orbital rings, or *shells* as they are more usually called, by the attractive force of the nucleus. The different shells are distinguished by assigning to them letters of the alphabet, starting at K for the innermost, and proceeding through L, M, N, etc., to the outermost. The electrons (or electron) making up the outermost shell are called *valence electrons*, and these being farthest from the attractive force of the nucleus are least tightly bound in the complete atomic assemblage. It is the valence electrons that play the active part in electrical conduction.

Figure 1.1 shows atoms of hydrogen and helium, both very light gases; and *Figure 1.2* shows the atoms of *germanium* and *silicon*, two very important elements in the manufacture of transistors. You should notice that each of these last two atoms has four valence electrons.

Hydrogen atom – a single proton around which revolves a single electron

Helium atom – a nucleus of two protons and two neutrons with two orbital electrons

Figure 1.1

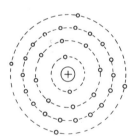

Germanium atom – a nucleus of 32 protons and 42 neutrons with 4 shells containing respectively 2, 8, 18 and 4 orbital electrons

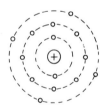

Silicon atom – a nucleus of 14 protons and 14 neutrons with 3 shells containing respectively 2, 8 and 4 orbital electrons

Figure 1.2

Bear in mind also that the diagrams are simply two-dimensional representations of the atoms and that in reality the orbiting electrons neither rotate in circles nor lie in one plane.

(1) The K, L, M and N shells of a copper atom contain, in order, 2, 8, 18 and 1 electron. How many protons are there in the nucleus?

(2) The atom of aluminium has a nucleus made up of 13 protons and 14 neutrons. In the two inner shells there are 2 and 8 electrons respectively. How many valence electrons are there in the aluminium atom?

CONDUCTION

In metals, and in the non-metallic element carbon, the valence electrons are easily displaced from their orbital shells, so that there is normally a great number of 'free' electrons wandering about within the material. These electrons haphazardly attach themselves to or detach themselves from atoms, or dance about in transit between the atoms. When a potential difference is set up across opposite ends of a piece of metal or carbon, this haphazard exchange is ended and the free electrons are constrained to drift on average towards the positive pole of the applied potential. This drift, which is a relatively slow affair, constitutes an electric current and the material in which the drift occurs is an electrical conductor. *Figure 1.3* shows the probable path of an electron in a piece of copper wire before and after the application of an external voltage.

Random movement of a free electron in a solid conductor

General drift of the electron towards a positive potential

Figure 1.3

Electrons are not necessarily the only free charge carriers. Complete atoms are electrically neutral so their motion does not constitute an electric current. Atoms which are not complete by virtue of having lost one or more of their electrons, however, will exhibit an overall positive charge and are known as positive *ions*. So ions are not electrically neutral and, like electrons, they can act as charge carriers. When a voltage is applied across the medium in which the ions are present, a drift will occur in the direction of the negative pole. It is easily possible for electrons and ions to be on the move at the same time. In a gas, for example, electrons and positive ions can be moving in opposite directions under the influence of an impressed voltage; if the velocity acquired by the electrons becomes great enough, they may readily knock other electrons away from their parent atoms, so creating further positive ions. If the production of free electrons and ions builds up sufficiently in this way, the gas emits a coloured glow and is said to be *ionised*. You have only to look at night-time advertising displays to see this effect in action.

In metals, however, the only charge carriers are electrons, for the atoms, even when they have lost one or more of their valence electrons, are fixed rigidly within the atomic structure of the material.

At this point we must make a note of some importance. Since the charge carriers in metal conductors are negatively charged electrons, the movement of the carriers is in the direction negative to positive, that is, *opposite* to the *conventional* direction of current which you will have used throughout your work in electrical principles. The true *electronic* flow of current becomes of significance in the study of electronic devices.

THERMIONIC EMISSION

Provided that the temperature is low, for example, ordinary room temperature, the free electrons in a metal drift only from atom to atom *within* the conductor. None escapes from the surface of the conductor into the surrounding space. This is because of the strong attractive force exerted upon electrons located at or close to the surface by the mass of atoms making up the conductor. The movement of free electrons may, however, be accelerated by the addition of heat energy. If the temperature of the metal is raised, the free electrons gain velocity and hence kinetic energy. When the metal becomes sufficiently hot, some electrons acquire enough energy to break away from the surface of the conductor. The rate at which such *thermionic* (heat induced) emission occurs from the heated surface depends upon the material concerned and the temperature.

The escape of electrons cannot continue indefinitely. As soon as the emission begins, other factors come into play which tend to restrain the loss of electrons:

1. When the conductor loses electrons from its surface, the surface becomes positively charged and this tends to attract the electrons back again.
2. Electrons which have already escaped form a negative barrier which discourages other electrons from following their example.
3. The air molecules surrounding the heated surface intercept the escaping electrons and absorb most of their surplus energy in the collision. The result is that the electrons fall back to the surface under its positive attraction.

The effect of thermionic emission is put to use by enclosing the heated conductor in an evacuated envelope, usually a glass bulb having the necessary connections brought out through air tight seals. In the very near vacuum which can be achieved in such a bulb the problem of collision with air molecules is eliminated and the electrons can evaporate from the heated surface without restriction of that kind. The back attraction of the positive surface and the formation of a negative region made up from the cluster of electrons immediately in the vicinity of the surface remain unaffected, however. The result is that a negative cloud, or *space charge*, forms close to the heated surface and remains there, provided there is no change in the temperature, at a constant density. The reason for this is not difficult to see. Those electrons at the outer edge of the cloud, having lost their kinetic energy, want to return to the heated surface but are prevented from doing so by the rest of the electrons between them and the surface. At the inner edge of the cloud, individual electrons continually leave the surface but are replaced by others which have used up their energy in doing work against the established cloud and return. Once established, therefore, the density of the cloud remains substantially constant. *Figure 1.4* illustrates the formation of a space charge. The heated conductor from which electrons are released in this way is called a *cathode*.

The phenomenon of thermionic emission is put to use in all electronic valves and cathode ray tubes.

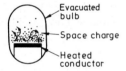

Figure 1.4

CATHODES

The cathode is an essential part of any electronic valve because it provides the electrons necessary for the operation of the valve. Although many metals when heated sufficiently will supply electrons in the manner already described, only a few such metals are of practical use. The materials in general use are tungsten, thoriated tungsten and metals which have been coated with alkaline earth oxides such as calcium, strontium or barium.

For a given material the rate of emission of electrons depends only upon the temperature, and if a graph of emission current i is plotted against temperature, the general shape of the curve obtained will be as shown in *Figure 1.5*. The emission is small until a certain temperature T_1 is reached, after which it increases very rapidly. The temperature at which a pure metal will emit enough electrons to be of practical significance is very high, and tungsten is one of the few metals which will provide a copious supply of electrons without itself melting under the effects of the high temperature, a dazzling white heat of something like 2200 °C. Thoriated tungsten emitters are made from tungsten

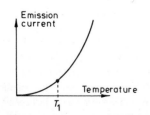

Figure 1.5

impregnated with thorium oxide. These emitters liberate electrons at a temperature of about 1700 °C, a bright yellow, and are much more economical of power than are the pure tungsten filaments. If alkaline earth oxides are applied as a coating to a nickel alloy base and dried, only a dull red heat of about 750 °C is sufficient to provide an abundant supply of emitted electrons. All the ordinary valves and cathode ray tubes you are likely to encounter will be operated from oxide coated cathodes.

The method of heating the cathode is used to distinguish between two different forms of cathode. The *directly heated* cathode is simply a wire mounted in an evacuated envelope and heated by the passage of an electric current (*Figure 1.6(a)*). The *indirectly heated* cathode consists of a thin walled metal sleeve or cylinder coated on the outside with alkaline earth oxide, while inside the sleeve is a coiled tungsten heating filament insulated electrically from the sleeve. This filament is used only for the purpose of heating the surrounding sleeve to the required temperature. It contributes nothing to the electron emission itself, which derives entirely from the coated surface of the sleeve. *Figure 1.6(b)* shows the construction. The assembly is, of course, mounted in an evacuated glass envelope.

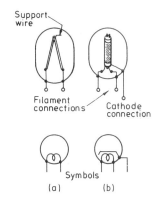

Figure 1.6

The manner in which the emitted electrons are put to work will be discussed in due course. Here is another problem to think about.

> (3) Would you agree that:
> (a) an oxide coated cathode must be of indirectly heated form?
> (b) a directly heated cathode has to be run considerably hotter than an indirectly heated one?
> (c) a space charge, once established, prevents the emission of further electrons?

SEMICONDUCTORS

Certain materials are in the class of electrical conductors, characterised by a high density of free electrons. All metals, together with carbon, come into this category. Other materials have virtually no free electrons available and these are classified as non-conductors or *insulators*. Rubber, glass and mica are examples. Between these extremes come the *semiconductors*, materials which, under ordinary conditions, are neither good conductors nor good insulators but can be made to exhibit some of the properties of each. Among the semiconductors having practical importance in modern electronics are foremost the elements germanium and silicon, with selenium, lead sulphide, copper oxide and cadmium sulphide also having general and specialised applications.

To understand the nature of a semiconductor it is necessary to look into the atomic arrangement of *crystalline* substances. We have already talked about the structure of individual atoms, but not about the way an assemblage of atoms group together to form larger bodies. Certain materials form themselves into bodies called crystals which have characteristic geometric shapes. Materials which do not take such forms are non-crystalline or *amorphous*. It is not necessarily possible to distinguish one from another simply by looking at it. A sheet of glass is amorphous, but a slab of quartz, which may be polished to resemble

6 *Thermionic and semiconductor theory*

glass, is crystalline. Crystal sizes vary enormously: a crystal of quartz takes the form of a hexagonal rod capped at each end by pyramids and may be many centimetres in length; a crystal of common table salt is a small rectangular block best seen under a good magnifying lens. Thin slabs or sections are cut from large crystals and put to a variety of electrical uses. Quartz crystals (incorrectly named), for example, are quartz *slices* used in microphones, pick-up heads and the like. So too, and of more concern to us, are wafers cut from germanium or silicon crystals and used in the manufacture of transistors and allied devices. Although carbon is a conductor, crystalline carbon is an insulator and a rather expensive one — diamond!

Inside a crystalline substance the outermost shell electrons (the valence electrons) of the individual atoms link up and arrange themselves with the valence electrons in adjacent atoms to form *co-valent bonds* which hold the atoms together in an orderly network or *lattice* structure. Thus, in any co-valent bond there are shared electrons, no atom having a monopoly in outermost electrons. Referring back to *Figure 1.2*, the arrangement of orbital electrons in a germanium atom in four shells is, reading from the inner K shell outwards, 2, 8, 18 and 4 electrons respectively; for the silicon atom the corresponding arrangement is 2, 8 and 4 electrons. It might appear that some or all of the four valence electrons in these atoms might easily be displaced from their orbits, to drift through the material as charge carriers under the influence of an applied voltage as they do in metals. In practice this does not happen; the valence electrons of each atom form co-valent bonds with neighbouring atoms and become very difficult to shift from their orbits. Crystals of pure germanium and pure silicon are therefore insulators, or at least extremely high value resistors. *Figure 1.7* is a two-dimensional representation of the effect of co-valent bonding. The valence electrons share themselves between four neighbouring atoms so that, in the case of germanium and silicon, the atoms behave as though each of their outer shells contain, not four, but eight electrons. In this condition the outer shell is in a stable state, there are no free electrons anywhere to act as charge carriers and the crystals are consequently electrical insulators. This situation is, of course, only true if the crystal structure is perfect and all the co-valent bonds are satisfied. There are always 'faults' and impurities present in the structure and these can provide free carriers so that perfect insulation is never possible.

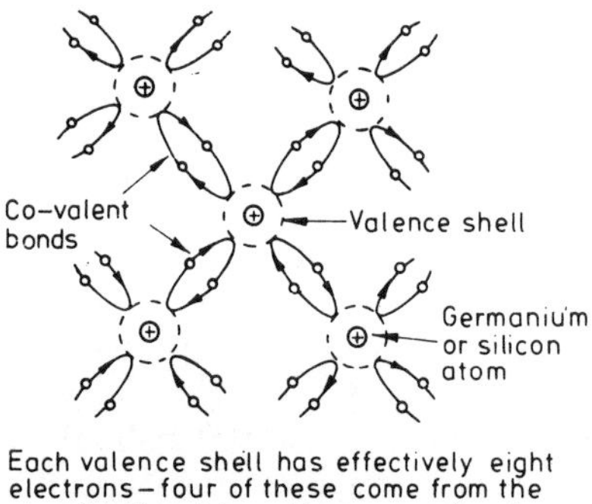

However, apart from this, if the temperature of germanium or silicon is raised, thermal agitation increases among the constituent atoms and some of the co-valent bonds are broken. Much of this goes on at ordinary room temperature, and the effect is accelerated as the temperature is raised above this level. This produces *thermally generated electrons* which act as negative charge carriers; so the crystal turns from a very good insulator into a poor conductor as its temperature is moderately raised. But the electrons are not the only charge carriers produced, unlike the case of thermionic emission from metals or ordinary electron movement in conducting wires. When a bond breaks and an electron is released, a vacancy or *hole* is left in the crystal structure. Since this hole has been formed by the removal of a negatively charged electron, the hole must have a positive charge associated with it. *Figure 1.8* illustrates the situation. The concept of positive hole charge carriers is, admittedly, a difficult one to appreciate, but you might consider it in this way: an electron is a basic

Each valence shell has effectively eight electrons—four of these come from the atom itself and four others come from four adjacent atoms

Figure 1.7

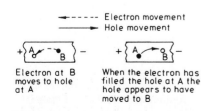

Figure 1.8

negative charge so that when it moves out of a valence bond it leaves behind a hole which then manifests itself as a net positive charge. Under the influence of an applied voltage, the now free electrons move towards the positive pole of the supply but on the way many of them fill the vacancies they find in the structure. In this way, referring to the diagram, a hole disappears at A by being filled, but effectively reappears at B, the point recently vacated by the electron which filled it. In other words, an electron moving out of its valence bond into an adjacent hole can be looked upon as being equivalent electrically to the hole itself moving in the opposite direction. So while electrons move towards the positive pole of the supply, holes effectively move towards the negative pole. Both hole and electron movement contribute to conduction. Thermally generated electrons and holes always appear in pairs, and the resulting conductivity of the crystal is called the *intrinsic conductivity*.

IMPURITY ATOMS

Intrinsic semiconductor material is of little practical importance because, as we have seen, the conductivity is very temperature sensitive and the process of conduction is due partly to electrons and partly to holes. The preparation of the material to overcome these drawbacks involves changes in the conduction characteristics so that either electrons *or* holes become the dominant charge carriers. This is done by mixing an extremely small quantity (about one part in 100 million) of a selected impurity into the semiconductor material. The atoms of this impurity material must have dimensions roughly equal to those of germanium or silicon atoms so that they will fit into the crystal lattice without seriously upsetting the regular geometric construction. When the atoms of the impurity material have more valence electrons than are required to satisfy the valence bonds with neighbouring semiconductor atoms, there will be electrons 'left over', and these will be free to participate in current conduction. Such an impurity therefore will give the semiconductor material an abundance of negative charge carriers; it is then referred to as n-*type material* and electrons are the *majority carriers*. *Figure 1.9* is a simplified representation of such an impurity atom in part of a semiconductor crystal lattice. Here, each semiconductor atom has four valence electrons and the impurity atom has five. Four of the valence electrons of the impurity link with those of neighbouring semiconductor atoms to form co-valent bonds and so complete the crystal structure. The extra electron is free to move through the lattice. An impurity of this type is called a *donor* since it donates electrons as charge carriers. Examples of *n*-type impurities are antimony, arsenic and phosphorus. The atoms of these elements are *pentavalent*, i.e. they have five valence electrons.

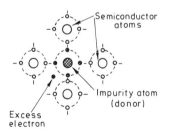

Figure 1.9

Suppose now that the atoms of the impurity material have less valence electrons than are needed to satisfy the valence bonds in the semiconductor crystal (*Figure 1.10*). As before, each semiconductor atom has four valence electrons but the impurity atom has only three. Only three valence bonds with adjacent atoms therefore are satisfied, so that a hole accordingly appears in what would have been the fourth bond. Such an impurity therefore will give the semiconductor material an abundance of holes as charge carriers; we now have p-*type material* and positive holes are the majority carriers. This time the impurity

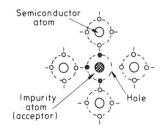

Figure 1.10

atoms are called *acceptors* because they can accept electrons from other atoms. Examples of *p*-type impurities are indium, aluminium and boron. The atoms of these elements are *trivalent*, that is, they have three valence electrons.

There is one point which you should be clear about at this stage. *n*- and *p*-type semiconductor materials have an electron 'excess' and an electron 'deficit' respectively, but the material is electrically neutral. You must not confuse the situation with electrostatic charges which can be built up on conductors or insulators; such charges represent a *displacement* of electrons from one part of a conductor to another, and such charges can be neutralised by contact with earth. Connecting a piece of *n*- or *p*-type material to earth will do nothing to remove the free electrons or fill the holes — if that happened there would be no such things as transistors!

One other thing before we turn to the practical applications of semiconductor material. Current moves more slowly through a semiconductor than in a true conductor. The electrons drift more slowly because they encounter obstructions due to imperfections in the crystal lattice, and the movement of holes is even slower than that of electrons. This may sound startling at first, but remember that while electrons may occupy any position within the lattice between holes, the holes themselves can only 'jump' from one valence bond to another — there are no intermediate positions at all.

Now here is your first set of problems. Not many to begin with, but although you are not expected to become atomic scientists, you will find some of the answers useful in your later work.

PROBLEMS FOR SECTION 1

(4) Complete the following statements:
   (a) The particles found in an atomic nucleus are ...... and ......
   (b) An electron is about ...... times 'lighter' than a proton.
   (c) Electrons in the outermost shells of atoms are ...... electrons.
   (d) Only the .... atom has no neutron in its nucleus.
   (e) The emission of electrons from a heated surface depends only upon the .... and the ....
   (f) The majority carriers in p-type material are ....
   (g) Increasing the temperature of a semiconductor crystal increases the .... and hence reduces the ....

(5) Say whether the following statements are true or false:
   (a) All indirectly heated cathodes are oxide coated.
   (b) The heating filament of an indirectly heated cathode may be fed from either a d.c. or an a.c. supply.
   (c) A space charge once established, remains at a constant density even if the temperature of the surface is increased.

(6) Fill in the missing numbers in the following table:

	Protons	K	L	M	shells
Aluminium	13	2		3	
Silicon	14	2	8		
Phosphorus			8	5	
Chlorine	17	2	8		

(7) The gallium atom has four shells and its nucleus contains 31 protons. The K, L and M shells contain respectively 2, 8 and 18 electrons. Would gallium impurity atoms produce *p*-type or *n*-type semiconductor material?

(8) Eight valency electrons surround a silicon or a .... atom in a crystal of semiconductor material .... of these come from this atom and .... others come from .... adjacent atoms. Fill in the missing words.

2 Semiconductor and thermionic diodes

Aims: At the end of this Unit section you should be able to:
Explain the behaviour and action of minority charge carriers.
Understand the operation of a p-n junction diode.
Compare junction potentials of germanium and silicon diodes.
Plot and describe the static characteristics of germanium and silicon diodes.
Explain the operation of a thermionic diode.
Plot and describe the anode characteristics of thermionic diodes.

We have mentioned that intrinsic conduction in a semiconductor crystal does not depend upon the addition of impurity atoms. Temperature rise itself is sufficient to break many of the co-valent bonds in the crystal lattice, with the result that electron-hole pairs are created, each capable of acting as a charge carrier. We say that both charge carriers are *mobile*. The semiconductor crystal does not then behave as an insulator, as it would if all the bonds were complete and satisfied, but becomes a conductor, though not necessarily a particularly good one. As temperature increases more free charge carriers are produced as equal numbers of holes and electrons and the conductivity of the material rises. Hence the resistance of the crystal falls as the temperature rises, which means that the material has a negative temperature coefficient of resistance.

When a semiconductor crystal has been *doped* with impurity atoms as described in the previous section, the conduction which results is said to be by *extrinsic* action. In *n*-type material, mobile negative carriers (electrons) are produced without corresponding mobile positive carriers (holes). In *p*-type material, mobile positive carriers are produced without corresponding mobile negative carriers. This does not mean that an increase in temperature no longer leads to intrinsic action within the lattice. The production of hole-electron pairs still goes on but the extrinsic production of charged mobile carriers completely swamps out the intrinsic action. In the *p*-type, the relatively few extra holes produced are insignificant compared with the enormous number of holes already present. However, the extra electrons produced play an important part in the conduction process as these are *n*-carriers in *p*-type material. These electrons (and similarly for the relatively few holes found for the same reason in *n*-type material) are known as *minority carriers*. Under an impressed voltage, minority carriers move in the opposite direction to majority carriers, but their number depends upon temperature, not the added impurity atoms.

Extrinsic conduction is illustrated in *Figure 2.1*. You must bear in mind that only electrons can act as carriers in the external connecting wires. The flow of electrons in the *n*-type material of *Figure 2.1(a)* should not present you with any difficulty in understanding, but you may not find the concept of hole flow in the circuit of *Figure 2.1(b)* quite so obvious. It will help if you recall that in this case the movement

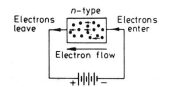

Figure 2.1(a)

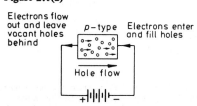

Figure 2.1(b)

of the holes through the crystal is equivalent to electrons moving in the opposite direction. Holes are filled (effectively disappear) at the right-hand side of the diagram as electrons enter from the negative pole of the battery; they surrender electrons (effectively appear) on the left-hand side of the diagram as the electrons return to the positive pole of the battery. In this way the holes move *only* through the semiconductor region of the circuit — their entrance on the left-hand side and exit on the right-hand side is all an illusion!

> (1) Does it make any difference to conduction which way round the batteries shown in *Figure 2.1* are connected?
> (2) Is the total current (as measured on an ammeter wired into the circuit) flowing through a piece of extrinsic semiconductor material less than, greater than or the same as the current would be if there were no minority carriers present?

THE JUNCTION DIODE

If a piece of *n*-type silicon (or germanium) and a piece of *p*-type silicon (or germanium) are alloyed into contact as shown in *Figure 2.2*, the unit is called a *p-n* junction and exhibits properties which enable it to be used as a 'one-way' device or diode rectifier. The word diode stems from the fact that there are two parts or electrodes in the assembly. At the junction, electrons from the *n*-type silicon tend to diffuse into the *p*-type silicon, and holes from the *p*-type silicon tend to diffuse into the *n*-type silicon. In the immediate neighbourhood of the junction, therefore, holes and electrons recombine. Once recombined in this way they can take no further part in the conduction process. This small initial diffusion* of holes and electrons across the junction sets up what is called a *depletion layer* or *potential barrier*; the depletion layer having no majority carriers within its boundaries behaves as an insulator. Once established, this insulating layer prevents any further migration of either electrons or holes across the junction. The condition is equivalent to a source of potential (and hence an electric field) acting across the junction and may be represented as an imaginary battery E connected as shown in *Figure 2.3*. The *p*-side has gained electrons, the *n*-side has gained holes; the *p*-side must therefore become negative with respect to the *n*-side. The polarity of the battery is then as indicated. Note this very carefully. The actual voltage of the battery depends upon whether the junction material is silicon or germanium. Its value is about 0.3 V for germanium and 0.7 V for silicon, the exact value depending upon the carrier densities.

Now suppose that connections are made at the ends of the *n*- and *p*-type materials and that a voltage is applied from an external battery as shown in *Figure 2.4*. When the *p* region of the junction is made positive, as in *Figure 2.4(a)*, the holes are repelled by the positive field and the electrons by the negative field. Both holes and electrons are driven in the direction indicated by the arrows, towards the *p-n* junction where they recombine. A high current flows, since the junction barrier is

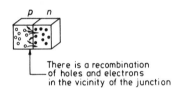

Figure 2.2

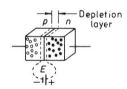

Figure 2.3

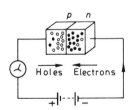

Figure 2.4(a)

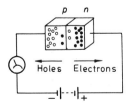

Figure 2.4(b)

---

*If you add a small amount of dye to a jar of water and set it aside without stirring or other disturbance, the dye will be found to have diffused through the water within a matter of a few hours. Diffusion tends to equalise concentration throughout a system.

effectively removed and the resistance is consequently low. We say that the junction is biased in the forward direction.

When the *p* region is made negative, as in *Figure 2.4(b)*, holes are attracted by the negative field and electrons by the positive field. Both holes and electrons are drawn away from the *p-n* junction, the depletion layer is effectively widened, and the junction resistance becomes very high. As a result of this action, the current flow is very low. We say that the junction is reverse-biased. A *p-n* junction therefore is a one-way device, permitting current to flow in only one direction.

But here we must add a qualification to the last statement. You will have noticed that when the junction is biased in the reverse direction, the current flow is stated to be low, not zero. Why is this? Look again at *Figure 2.4.* In both diagrams the majority carriers are found in their respective halves of the junction: holes (open circles) and electrons (filled circles). But you will notice that a few electrons have been drawn in the *p*-regions, and a few holes in the *n*-regions. These are the minority carriers, generated remember, by thermal agitation of the co-valent bonds. When the junction is forward-biased as at (a) these minority carriers simply move along with the majority carriers and do nothing more than make a very small contribution to the relatively large current which is flowing in the circuit. When the junction is reverse-biased, however, as at (b), only the majority carriers are drawn away from the junction; the applied polarity is such that the minority carriers are attracted towards the junction. Hence a current flows in the circuit due entirely to the presence of the minority carriers. This current, which is normally extremely small, is called the *leakage current.* Notice that its direction is opposite to that of the large forward current which flows under forward bias conditions.

> (3) If a junction diode is raised in temperature, what happens to (i) the forward current, (ii) the reverse leakage current — if anything?

You will have looked up the answer to the previous problem — after attempting it, of course. Raising the temperature causes a rapid increase in the generation of minority carriers and hence the leakage current. At around room temperature each increase of 10 °C roughly doubles the rate of generation for germanium, or of 5 °C for silicon. This might make it seem that germanium would be the better material to use where high ambient temperatures were concerned, but this is not so. Although the rate of increase is greater for silicon, its actual value at room temperature is considerably less than that of germanium — so silicon is used where high temperatures are likely to be encountered.

We require now a symbol for a semiconductor diode, and this is illustrated in *Figure 2.5.* When you use this symbol, keep in mind that the arrow points in the direction of the conventional flow of current through the diode. On this basis, the diode will conduct if the arrow is connected to the positive pole of the battery; it will not conduct if the arrow is connected to the negative pole of the battery. Many diodes look like small resistors and they are nearly always marked with a spot or coloured ring at one end. Such a marked end corresponds to the line

Figure 2.5

Characteristic Curves

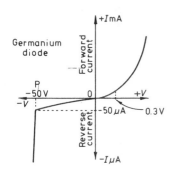

Figure 2.6(a)

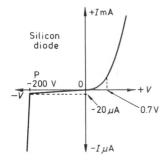

Figure 2.6(b)

in the symbol, not the arrow, so this end has to be negative for the diode to conduct.

If the temperature remains substantially constant, the only factor controlling the flow of current through a semiconductor diode is the applied voltage. If applied voltage and current are measured in a series of steps, the curve which results when the measured quantities are plotted as a graph is called the *static characteristic*. Static characteristic curves for germanium and silicon diodes are shown in *Figure 2.6* at (*a*) and (*b*) respectively. Forward voltage is plotted along the horizontal axes to the right of the origin and forward current is plotted vertically above the origin. This is the region in which the diodes conduct. Notice that the forward current does not become significant until the applied voltage is roughly equal to the barrier voltage set up at the depletion layer, about 0.3 V for germanium and 0.7 V for silicon, as already mentioned. After this, current rises very rapidly as the applied voltage increases. Clearly, the voltage cannot be increased indefinitely and the current cannot rise indefinitely; if this is attempted, the diode will be destroyed.

The part of the curve to the left of and below the origin is the characteristic for a reverse-biased condition. The applied p.d. is now reversed ($-V$) and the current which flows is also reversed. This is simply the leakage current due to the minority carriers. Note the scaling of the vertical axes: forward current is in milliamperes, while reverse current is in microamperes. This scale change is necessary to show the curve of reverse current in detail. In the reverse direction the leakage current in the germanium diode tends to increase with voltage, whereas the silicon diode current is not only generally much smaller but tends to remain fairly constant. A point of particular significance appears at P on each curve, typically at about -50 V for germanium and -200 V for silicon, when the reverse current suddenly increases and builds up very rapidly to a high value. This effect is known as *avalanche breakdown* and unless the reverse current is limited in some way, results in the destruction of the diode. The reason for the avalanche breakdown is that if the reverse voltage becomes too great, the minority carriers are accelerated to the point where (a) they begin to heat up the diode, and (b) they collide with atoms in the depletion layer and dislodge further electrons, so creating further hole-electron pairs and hence more minority carriers. The effect 'avalanches' and rapidly results in a destructive flow of current.

> (4) The leakage current of a certain germanium diode is 5 μA at 20 °C. What order of leakage current would you expect at 80 °C?
>
> (5) Does a change in ambient temperature affect the value of reverse voltage at which a semiconductor diode will go into avalanche breakdown?

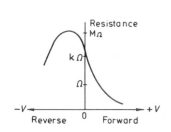

Figure 2.7

Figure 2.7 shows the static resistance characteristic of a semiconductor diode. As the forward voltage is increased, the resistance of the diode falls to a very low value. At decreasing values of forward voltage the resistance increases until, just above zero voltage, the depletion layer appears and the resistance becomes very high. As zero voltage is passed and reverse voltage is increased, the resistance reaches a peak value, and then decreases as the minority carriers begin to assume significant proportions.

> (6) Sketch a circuit suitable for plotting the static voltage-current and resistance curves of a semiconductor diode.

THE THERMIONIC DIODE

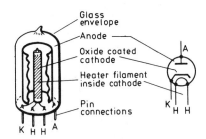

Figure 2.8

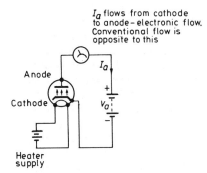

Figure 2.9

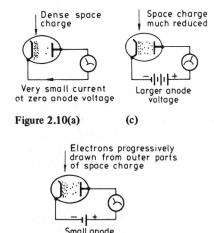

Figure 2.10(a)   (c)

Figure 2.10(b)

The thermionic diode, or two-electrode valve, operates as a one-way device as does the *p-n* semiconductor diode. In appearance and in its mode of working it is completely different from the *p-n* diode. It consists of an emitter (the cathode) which in nearly all instances is in the form of a cylinder or sleeve, oxide coated and heated from an internal filament; and a collecting electrode (the anode) which takes the form of another cylinder axially surrounding the cathode sleeve. A general sketch of the construction is given in *Figure 2.8*, which shows also the symbol for a thermionic diode. The whole assembly is rigidly supported in an evacuated glass envelope.

A low voltage source is used to heat the filament which in turn heats the cathode sleeve to the proper operating temperature. Apart from this essential function, the heater filament is of no further interest. Electrons are emitted from the cathode and if the anode is maintained at a positive potential relative to the cathode, a current will flow between them and hence around the external parts of the circuit as indicated in *Figure 2.9*. Actually it will be found that a very small current flows even when the anode voltage (V_a) is zero. Recall that a space charge forms around the cathode and a highly negative region exists close to the cathode surface because of this. A few electrons on the outermost boundaries of the space charge are repelled strongly by this intensely negative field and find their way to the anode which is itself, with respect to the space charge, very slightly positive. These electrons then return to the cathode by way of the external circuit. *Figure 2.10(a)* shows this situation.

When the anode is made slightly positive, as in *Figure 2.10(b)*, electrons begin to be drawn from the outer fringes of the space charge, though at this stage no further electrons will be accepted from the cathode itself by the space charge. A negative region will still exist in front of the cathode face, even though the anode is now positive. The external current now increases accordingly. As the anode voltage is further increased (*Figure 2.10(c)*) the anode current will further increase, more and more electrons being drawn from the space charge. At the same time the space charge will reduce in density and effectively move towards the cathode, since the electrons closest to the anode are being progressively removed. All the time the space charge exists, the anode current (for a given anode voltage) is limited by what can be drawn from the outermost fringes of the charge, and the anode current is said to be space charge limited. It is important to appreciate this point. The number of electrons in transit at any instant is just sufficient to produce a negative space charge which shields from the attraction of the anode any electrons just then leaving the cathode. Anode current I_a depends therefore only upon V_a, being quite independent of the emission rate at the cathode.

As the anode voltage is increased still further, a point is reached

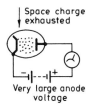

Figure 2.10(d)

when the rise in anode current slackens and there is no further increase in the current even though the anode voltage may be advanced indefinitely. In this state all the electrons emitted by the cathode are being collected by the anode, the space charge now having ceased to exist. I_a has now reached its maximum possible value for the particular cathode temperature established. The diode is then said to be saturated or temperature limited. *Figure 2.10(d)* shows the situation where the emitted electrons move directly to the anode without spending some of their time in an intervening space charge.

Like the *p-n* junction diode, the thermionic diode conducts only in one direction: when the anode is positive with respect to the cathode. If the polarity is reversed, no anode current is possible.

Characteristic Curves

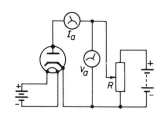

Figure 2.11

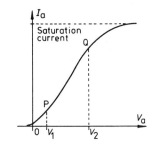

Figure 2.12

A circuit suitable for obtaining a series of related values of V_a and I_a is sketched in *Figure 2.11*. The potentiometer R is advanced in steps of, say, 5 V, from zero to a maximum dictated by the particular diode, and readings of I_a are recorded at each step. A graph then can be plotted showing the variation in I_a as V_a is advanced. Such a curve is drawn in *Figure 2.12*. This figure shows that between the points marked P and Q the characteristic is substantially straight. This being so, the diode is behaving for anode voltages between the limits V_1 and V_2 as a linear resistance, I_a being directly proportional to V_a. Beyond Q the curve bends over as saturation sets in. The small initial bend between the origin and point P is exaggerated for clarity; it is usually much less marked than the figure indicates. No reverse voltage characteristic is given as the current remains at zero for all anode voltages below zero.

For the correct cathode temperature, i.e. for the heater voltage set to the manufacturer's specification, it is not normally possible to run the diode into the saturation region of the characteristic, and diodes are never used in this region. It is, however, possible to obtain saturation region characteristics experimentally by deliberately reducing the heater voltage and hence the cathode temperature. The circuit of *Figure 2.11* can be used for this, with the addition of a suitable variable resistor of a few ohms value connected in series with the heater. By underrunning the heater by 50% and 75% of its normal operating voltage, the cathode emission is reduced sufficiently for saturation to begin at relatively low values of anode voltage, and in this way it is possible to obtain curves similar to those shown in *Figure 2.13*.

COMPARISONS

Both the semiconductor and the thermionic diode do the same electrical job — they perform as one-way devices. Once the thermionic diode reigned supreme, but it has now almost universally been replaced by the semiconductor. This is because the semiconductor diode requires no heater supply so there is no wasted electrical and thermal energy, and therefore no ventilation problems. Also it is very compact, robust, and can be made to handle large currents without excessive voltage loss because of its very low forward resistance. On the other hand it has the problem of leakage current when reverse-biased, so that the reverse resistance is not infinitely great (as it is in the thermionic diode), and reverse breakdown occurs at a much lower voltage than it does in the thermionic diode.

In the next Unit section we shall look into the applications of diodes. Before that, here are some test problems.

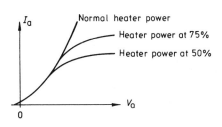

Figure 2.13

PROBLEMS FOR SECTION 2

Group 1

(7) Explain the meanings of the following terms: (i) valence electron, (ii) hole, (ii)

(7) Explain the meanings of the following terms: (i) valence electron, (ii) hole, (iii) impurity atom, (iv) leakage current, (v) avalanche breakdown.

(8) Fill in the missing words:
 (a) The space charge is densest near the ....
 (b) A thermionic diode biased in the forward direction should be operated only in the .... region.
 (c) Comparing germanium and silicon diodes, the .... voltage is higher but the .... current is lower in the silicon diode.
 (d) For a given cathode material, thermionic emission depends only upon the ....
 (e) When a semiconductor diode is forward biased, the leakage current flows in the .... direction .... the main current.

(9) Are the following statements true or false:
 (a) A semiconductor diode has two junctions.
 (b) In a semiconductor diode, it is the majority carriers that diffuse across the junction.
 (c) Leakage current in the circuit wires connected to a semiconductor diode consists only of holes.
 (d) The reverse current in a germanium diode is very much greater than in a silicon diode.
 (e) In p-type material a large number of holes = a small number of electrons + a large number of negatively ionised acceptor atoms.

(10) The current in a thermionic diode is first space charge limited and then temperature limited. Explain the meaning of this statement.

(11) The leakage current in a certain silicon diode is 0.05 μA at 25 °C. What will be its approximate value at 100 °C?

(12) After the method illustrated in *Figure 2.7*, sketch a resistance/voltage curve for a thermionic diode.

(13) What properties would you expect a perfect diode to exhibit? If there was such a device as a perfect diode, what would its characteristic curve look like? Make a sketch.

(14) Which electrode, anode or cathode, of a thermionic diode corresponds to the n-region of a p-n junction diode?

(15) The performance of a semiconductor diode is affected by increases in the ambient temperature. Do you think this statement is also true of thermionic diodes?

(16) Sketch a typical characteristic curve for a p-n junction diode, and explain its appearance from the movement of the carriers concerned.

Group 2

(17) The two diodes shown in *Figure 2.14* can be assumed to have zero forward resistance and infinite reverse resistance. Calculate the currents in each branch of the circuit (i) with the battery connected as shown, (ii) with the battery reversed.

(18) The width of the depletion layer (which may be assumed to be a perfect insulator) depends upon the applied reverse

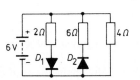

Figure 2.14

voltage. Can you think of any practical application to which this effect might be put?

(19) At the avalanche breakdown point of a semiconductor diode, the voltage across the diode $(-V)$ is practically constant, irrespective of the current flowing through the diode. Look back at *Figure 2.6*. Ignoring the possibility of burn-out for the moment, can you think of any practical application of this effect?

(20) A thermionic diode has the following characteristics

V_a (V)	0	10	20	30	40	50	60	70	80	90	100
I_a (mA)	0	1.0	2.5	4.1	6.0	8.2	10.4	12.9	15.3	18.4	21.7

Plot the characteristic curve neatly on graph paper. This diode is connected in series with a 500 Ω resistor and a d.c. supply (forward biased). Find (i) the current flowing, (ii) the p.d. across the circuit, when the p.d. across the diode is 65 V.

(21) A voltage increases uniformly from zero to 100 V in a time of 1 s. This voltage is applied to terminals A and B of the circuit shown in *Figure 2.15*. Assuming that the diode has a constant forward resistance of 50 Ω and an infinite reverse resistance, sketch the voltage waveform you would expect to obtain at terminals C and D during the 1 s rise.

(22) A thermionic diode is connected to a 50 V battery, anode positive, and a current of 10 mA flows in the circuit. Where is the power represented by the product of voltage and current being dissipated?

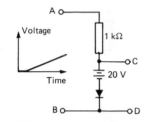

Figure 2.15

3 Applications of semiconductor diodes

Aims: At the end of this Unit section you should be able to:
Discuss the basic application of semiconductor diodes to power rectifier circuits.
Understand the action of a Zener diode as a voltage stabiliser and overload protector.
Understand the operation of a Varactor diode.
Explain the meaning and significance of peak inverse voltage.
Explain the purpose and operation of simple smoothing circuits.

We have now dealt in some detail with the theory of both thermionic and semiconductor diodes. Some of the more common applications of diodes will now be considered, and in Unit Section 9 further applications in the field of logic circuits will be discussed. In the applications which follow, where non-specialised diode elements are concerned, bear in mind that the circuit actions described are exactly the same irrespective of whether thermionic or semiconductor diodes are being employed. We shall, however, deal throughout with semiconductor elements only, as the thermionic diode is now little used in electronics.

RECTIFIER CIRCUITS

A rectifier is a device by which a direct current can be obtained from an alternating voltage supply, and the process of such a conversion is called *rectification*.

As you will have learned from your Electrical Principles programme, an alternating current is one which flows first in one direction along a conductor and then, after a given time, reverses and flows for a further period of time in the opposite direction. As we recall, such a complete backwards and forwards movement of the charge carriers (electrons only in this case) constitutes one cycle of the alternating current. *Figure 3.1* shows examples of alternating currents. On the other hand,

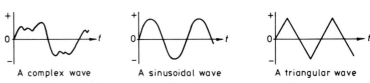

These are all examples of alternating electrical quantities

Figure 3.1

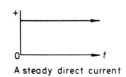

Figure 3.2(a)

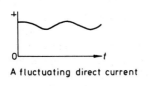

a direct current is one which flows always in one definite direction, though it is not necessarily a steady or constant current. A current flowing from the positive to the negative terminal of a well charged battery through some form of resistance or load is generally a steady flow, maintaining a certain level for the whole length of time that the circuit is operating. The curve of such a current is shown in *Figure 3.2(a)*. In (b) a fluctuating direct current is illustrated. At first glance this might be confused with an alternating current, but in this diagram the current never crosses the horizontal axis. Electrons in this circuit move in a cyclic manner, but they never turn around and move backwards.

Almost all electronic equipment operates from steady direct current supplies. In many instances such supplies are obtained from cells or batteries: portable radios, pocket calculators and flashlamps are familiar examples. But we do not normally operate our television receivers from battery supplies, and certainly it would not be very practicable to operate, say, large computer systems or high power transmitters from batteries. We use the mains electricity supply for such purposes, hence it becomes necessary to convert the alternating current (or voltage) of this supply into the direct current (or voltage) that our apparatus requires.

The simplest type of rectifier is a circuit element which performs the function of an automatic switch. A perfect rectifier would have zero resistance one way and infinite resistance the other, and the rectifier circuits we are going to examine will be assumed to have perfect rectifier elements. Clearly, our one-way *p-n* junction diode will serve the purpose of a rectifier element.

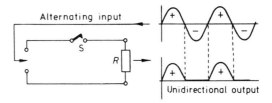

Figure 3.3

To convert the sinusoidal alternating current flowing in resistor R of *Figure 3.3* into a one-way or unidirectional current it is necessary to eliminate one half or other of the alternations. The curve must lie wholly above the horizontal axis or wholly below it, but it must not cross it. The obvious solution is to switch the circuit off whenever the current is about to reverse. If switch S is closed for each positive half-cycle of input current and opened for each negative half-cycle, the resulting current flow through R will be a series of positive pulsations, that is, a current which continually goes on and off but never reverses in direction. It is true we are using only one half of each input cycle and the unidirectional current we are getting as our output is a long way from being the steady current we require, but the method nevertheless provides us with a starting point from which we can now develop a number of rectifier systems.

THE HALF-WAVE RECTIFIER

The circuit of a half-wave rectifier is shown in *Figure 3.4*. This includes a transformer (as will all the circuits to be discussed) which is used for two basic reasons: it isolates the equipment being supplied from direct connection to the mains supply and so increases safety, and it enables the mains voltage to be either increased or decreased to a level suited to the apparatus for which the rectified supply will be finally required. It is not directly associated with the process of rectification which is our present concern. We are interested only in the fact that an alternating voltage is present at the secondary terminals A and B.

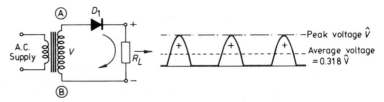

Figure 3.4

When terminal A is positive with respect to B, the diode conducts. The positive half-cycle of voltage across A–B therefore causes a current to flow around the circuit and a voltage will be developed across the load resistor R_L corresponding to the form of the half-cycle wave. When the input polarity reverses, terminal A will be negative with respect to B and the diode will switch off.

The voltage developed across R_L thus consists of half-sinewaves, and the circuit is known as a *half-wave rectifier*. The current through R_L is always in one direction, as shown, in spite of the on-off fluctuations; so our output is unidirectional. Notice that the peak value of the output wave is equal to the peak value of the alternating input voltage from the transformer, $\hat{V}$. You will also recall that the average value of the output wave is $\hat{V}/\pi$ or about $0.318\,\hat{V}$, shown in the broken line in *Figure 3.4*.

THE FULL-WAVE RECTIFIER

The half-wave rectifier has the disadvantage that there is no output at all for half of the available input. The full-wave rectifier, as its name implies, enables us to use both half-cycles of the input wave. The circuit is shown in *Figure 3.5*. Two diodes are used now, together with a transformer whose secondary winding is centre tapped at C. We can treat the centre tap as being a neutral point so that terminals A and B swing alternately positive and negative about it. Each diode consequently conducts in turn when its particular anode happens to be positive with respect to the centre point C. What we really have is two half-wave rectifier circuits connected to a single load resistor R_L, and each of these 'part' circuits take it in turn to supply current to the load. Following the direction of current flow as the diagram indicates, we notice that the current through R_L is, for both diode circuits, in the same direction. The output voltage developed across R_L is therefore unidirectional, the spaces between the half-sinewaves developed by either diode section now being filled in by the other diode section. The average output voltage across the load is now double that of the half-wave rectifier circuit, that is, about $0.636\,\hat{V}$. The peak voltage of the output wave is, as before, equal to the peak value of the alternating input voltage from *each half* of the secondary winding. Having to use the double winding on the transformer makes this component more bulky in size and of course more expensive.

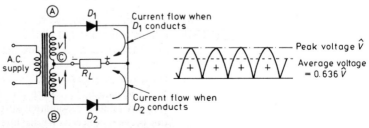

Figure 3.5

THE BRIDGE RECTIFIER

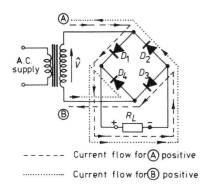

- - - - - Current flow for (A) positive
·········· Current flow for (B) positive

Figure 3.6

This form of full-wave rectifier uses four diodes but does not require a centre tapped transformer. As diodes are much cheaper components than transformers, this circuit is less expensive and often less bulky than the two-diode full-wave rectifier.

The circuit of the bridge rectifier is shown in *Figure 3.6*. This time the diodes conduct in series-pairs. When secondary terminal A is positive with respect to B, diodes D_1 and D_3 conduct in series, but diodes D_2 and D_4 are switched off. When the input polarity reverses, D_2 and D_4 switch on in series, but D_1 and D_3 switch off. Following the current direction through in each case, you will see that the current flow through load R_L is always in one direction as indicated. Once again a unidirectional current is obtained and so the voltage developed across R is unidirectional. As for the two-diode full-wave rectifier, the average output voltage is $0.636\ \hat{V}$.

Go through the above description carefully and make sure you understand the working of the bridge rectifier, as it has many applications in electronics.

PRACTICAL DIODE RATINGS

Certain voltage and current ratings are of importance in connection with rectifiers, and now that we have dealt with the three rectifier circuits commonly found, we shall illustrate these rating factors by referring them to the circuits concerned.

Power Dissipation

The ideal or perfect diode element we have so far assumed does not dissipate any power. In the forward direction its resistance is zero, hence whatever current flows through it, the voltage developed across it must be zero. In the reverse direction its resistance is infinite, hence whatever voltage is developed across it, the current flowing through it must be zero. In both cases, the power dissipated is zero. In the real-life diode, however, the forward resistance is never zero and the reverse resistance is never infinite, so power must be dissipated within the element.

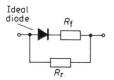

Figure 3.7

An *equivalent circuit* for a practical diode can be obtained by assuming that we have a perfect element, but connected in series with this is a small value resistor R_f representing the actual forward resistance, and connected in parallel is a large value resistor R_r representing the reverse resistance. The circuit is shown in *Figure 3.7*. You will encounter equivalent circuit representation of this sort throughout your electronics programme. When a voltage is applied in the forward direction to this circuit the apparent resistance of the device is very closely equal to R_f, while reverse voltage can send current only through R_r, the ideal element itself cutting off R_f completely. So power is dissipated in these resistive components of the real semiconductor diode in both the forward and the reverse directions of applied voltage. This dissipation will appear in the form of heat at the junction. It is necessary to ensure that this local heating does not lead to an appreciable rise in temperature and so to an appreciable increase in the leakage current across the junction. Failure to do this may well lead to the destructive 'avalanche' effect mentioned in the previous Unit section, even though the diode is nowhere close to its reverse voltage breakdown point. For this reason, rectifier diodes intended for use in high current circuits are made with thick based metal cases provided with fixing studs which enable them to be bolted down to large area metal plates which then act as so-called *heat sinks* and

Peak Inverse Voltage

permit the heat to be rapidly conducted away from the junction. A typical high power rectifier diode element is shown in *Figure 3.8*.

There are two important and related factors which have to be considered in the design of rectifier circuits in addition to the problem of power dissipation: these are the reverse breakdown point of the diode, and the *peak* operational voltage present across the diode when it is non-conducting. A rectifier diode must, of course, operate well away from its breakdown point, and here the main objective of the diode manufacturer is to ensure that breakdown occurs at far greater reverse voltages than the device will normally be subjected to in use. A rectifier diode must therefore be used only in a circuit where the applied reverse voltage is never of sufficient amplitude to approach the breakdown point. The value of the breakdown voltage for a particular diode is quoted by the manufacturer under the heading of the *peak inverse voltage* (p.i.v.).

Peak inverse voltage is the maximum voltage appearing across the terminals of a rectifier and acting in the reverse direction. The figure quoted for the p.i.v. usually implies that it represents the maximum reverse voltage that may be applied to the diode without reverse breakdown occurring, and this is the sense in which we will use the term.

We shall reconsider the three rectifier circuits already covered in terms of the peak inverse voltages present across the diodes, and for convenience the diagrams are reproduced in skeleton form in *Figure 3.9*.

In the case of the half-wave rectifier at (*a*), when the diode is reverse biased and switched off, no current flows in R_L, hence the applied voltage appears solely across the diode. The p.i.v. is therefore equal to $\hat{V}$, the peak value of the transformer secondary output. This means that the diode breakdown voltage must be greater in this case than $\hat{V}$, *not* the r.m.s. value of voltage in which the transformer output will be normally stated.

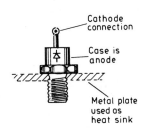

Figure 3.8

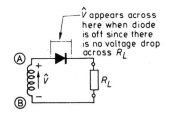

Arrows indicate instantaneous direction of voltage

Figure 3.9(a)

> (1) A diode has a p.i.v. rating of 300 V. Could this diode be used as a half-wave rectifier with a transformer having a stated secondary voltage of 250 V?

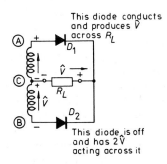

Figure 3.9(b)

Consider now the full-wave circuit shown at (*b*). At the instant when terminal A is at its maximum positive swing, diode D_1 will be fully conducting and the voltage across the load resistor R_L will be $\hat{V}$ with the polarity indicated. The voltage between points C and B will also be $\hat{V}$ and C will be positive with respect to B. Hence, as far as diode D_2 (which is reverse biased at this time) is concerned, it has across its terminals the load voltage in series with the voltage across CB. These two voltages are aiding each other, hence they combine to give a total reverse voltage across D_2 of $2\hat{V}$. The p.i.v. in this case is therefore twice the peak voltage on either half of the transformer secondary. As the full-wave two-diode rectifier is in very common use, this aspect of the p.i.v. present on the diodes must be carefully kept in mind.

> (2) A mains transformer is rated as giving an output of 350 V on each half of its secondary winding. What will be the p.i.v. across the diodes used in a full-wave circuit with this transformer?

Applications of semiconductor diodes 23

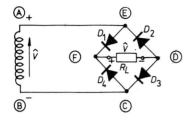

Figure 3.9(c)

Turning now to the bridge rectifier shown in (c), we take again the instant when the applied voltage is at its peak value and terminal A is positive. The voltage across the load resistor will be $\hat{V}$ with the polarity shown, diodes D_1 and D_3 now conducting. If we follow around the circuit AEDFCB through series diodes D_2 and D_4 which are now reverse biased and switched off, we see that the load voltage adds to the transformer voltage to give (as in the previous case) a total of $2\hat{V}$. This time, however, the voltage is shared between diodes D_2 and D_4 in series, hence the p.i.v. per diode is simply $\hat{V}$.

SMOOTHING CIRCUITS   The pulsating output we obtain from the rectifier circuits so far discussed is not suitable as it stands for the operation of equipment which requires a steady d.c. supply such as we should obtain from batteries. To iron out the rough outputs, as it were, smoothing circuit networks are connected to the rectifiers. The most simple case is illustrated in *Figure 3.10*, which shows a half-wave rectifier with a *reservoir capacitor*

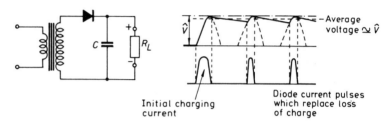

Figure 3.10

connected across the load. This capacitor is of large capacity, generally within the range 10–100 μF, or even greater. Consider what happens during the first positive output half-cycle after switching on. The capacitor will charge up to the peak value of the rectifier output voltage, that is, to $\hat{V}$. During the interval represented by the missing negative half-cycle of output no further charge is added to C and the voltage across its terminals falls slightly as it discharges through R_L. However, provided the discharge is relatively slow, which in effect is the same thing as saying that the value of R_L is large, the fall in terminal voltage before the arrival of the next positive half-cycle is very small. The capacitor is then 'topped-up' by the tip of this half-cycle, and the process then continues for the whole of the time that the circuit is switched on. You will notice two important points: (a) the waveform shown in full line is very much smoother than it was before the capacitor was added, (b) the average voltage across the load resistor is much greater. The best smoothing effect is obtained when the value of R_L is extremely high, for then there is negligible discharge between the output half-cycles and the almost steady d.c. output approximates in amplitude to $\hat{V}$. When the load resistor is small, however, the discharge becomes large and the output voltage is again rippled, though not to such a severe extent as that of an unsmoothed supply. The greater the capacity of the capacitor, for a given value of R_L, the better the smoothing. The product of C and R_L is called the *time constant* of the circuit.

One further important point must now be discussed. Once the capacitor is charged, the output voltage is, on average, approximately equal to $\hat{V}$. Suppose that it is equal to $0.9\,\hat{V}$. This means that it is providing a voltage which is acting in opposition to the voltage provided by the transformer secondary winding on the positive half-cycles during which the diode normally conducts. The diode cannot conduct, however, until its anode voltage becomes greater than the voltage on its cathode. If the voltage on the cathode is, for our illustrative example, $0.9\hat{V}$, then the diode remains switched off until the positive swing at transformer terminal A exceeds this value. The diode, therefore, now conducts over only a small part of each positive half-cycle from the transformer, whereas without the inclusion of capacitor C, it conducted over each complete half-cycle. The lower graph of *Figure 3.10* shows the corresponding pulsations of diode current. There are two dangers here for the unwary builder of a rectifier unit with added smoothing. The period during which the diode is conducting is a very small fraction of a cycle, and during this time it has to supply to the capacitor all the charge lost during the remainder of the cycle. The surge of current can therefore be very large, and hence the power dissipation in the diode can be considerable, even though its forward resistance may be small. Further, since the capacitor retains most of its charge between cycles, the voltage across the load resistor adds to the inverse voltage of the supply during the negative half-cycle when the diode is reverse biased. The p.i.v. in a half-wave rectifier is therefore almost equal to twice the peak transformer voltage, i.e. double the value it reaches in the absence of the smoothing capacitor. The diode chosen for the job must be rated to withstand this.

> (3) A diode has a forward resistance of R_f ohms and a reverse resistance which may be considered infinite. It is used in a circuit similar to that shown in *Figure 3.4*. Sketch a graph of the output variations of current through the load resistor R_L. If the transformer peak output voltage is $\hat{V}$, write down expressions for (a) the average current, (b) the r.m.s. current in the load*.

THE ZENER DIODE

You will recall that under reverse bias conditions, the only current flow in a *p-n* diode circuit is that due to minority carriers passing across the depletion layer. In an ideal diode this current would be zero, but in a real diode it is present though normally very small at ordinary temperatures: a matter of a few microamperes for a germanium diode and a mere nanoampere (10^{-9} A) or so for a silicon element. As the reverse bias is increased there is little effect on the flow of minority carriers, particularly in silicon diodes, and the typical characteristic shown in *Figure 3.11* shows that an almost constant reverse current flows. However, if the reverse voltage is continually increased, the point of breakdown is eventually reached and the current increases very suddenly.

We can now examine in a little more detail the mechanism of the breakdown or avalanche effect. *Figure 3.12* shows part of the crystal lattice of the semiconductor material, with four atomic nuclei each with its four valence electrons. With a sufficiently high reverse voltage, an electron E_1 is accelerated through the lattice with enough velocity to dislodge a valence electron E_2. This electron now comes under the

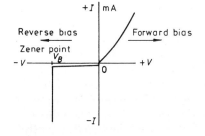

Figure 3.11

---

*We are dealing with a non-sinusoidal waveform when we consider current. It can be proved that the r.m.s. value of the rectified waveform current is $\hat{I}/2$.

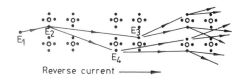

Figure 3.12

influence of the applied field and is similarly accelerated. In turn, these two electrons collide with further valence electrons E_3 and E_4 and break the co-valent bonds. There are now four electrons to continue the effect of dislodging further electrons and so build up what quickly becomes an avalanche of electrons throughout the lattice. If the breakdown is uncontrolled, the diode will be destroyed in an extremely short time, but damage can be prevented by connecting a resistance of a suitable value in series with the diode. At the point of breakdown, the maximum avalanche current can then never exceed the current determined by the resistor, and this current can be limited to a value which makes overheating and damage to the diode impossible.

Now it might seem that the best thing to do in any design incorporating semiconductor diodes would be to make sure (as we have emphasised for the rectifier circuits) that the reverse voltage was never great enough to get anywhere near the breakdown point. This is true for diodes used as rectifiers, but there are other diode applications which actually make use of the avalanche condition. Such diodes are known as *Zener diodes* or *voltage regulator diodes*. These diodes are deliberately connected into a circuit in the reverse direction, that is, the cathode terminal is connected to the positive pole of the supply and the anode to the negative pole. The diode is then reverse biased and avalanche is possible as soon as the applied voltage reaches an appropriate level.

Look again at *Figure 3.11*. Notice that after breakdown, the voltage across the diode remains virtually constant at the level V_B in spite of the increasing current flowing through it. This means that provided the avalanche current is prevented from becoming destructively great, we can use such a device as a source of constant voltage. Such Zener diodes are available with a wide range of breakdown voltages, from about -2 V down to some -200 V, the actual characteristic being determined by the impurity concentrations introduced during manufacture. They are also made in a range of power handling capabilities from half a watt or so up to many watts. The symbol for a Zener diode is shown in *Figure 3.13*.

Zener diode symbol

Figure 3.13

We shall now look at one or two applications of the Zener diode.

THE ZENER STABILISER

The circuit requirements of a simple Zener diode voltage stabiliser are remarkably simple: apart from the diode itself, which is a very inexpensive device, only one resistor is necessary. If we cast our minds back to the rectifier and smoothing circuits discussed earlier, it is clear that some ripple always remains on the output voltage, and that this becomes more pronounced as the current drawn from the reservoir capacitor by the load increases. This in turn reduces the average output voltage. In cases where the current drawn from the power unit is substantially constant there is no particular problem, but where the current is liable to wide variation, such as additional loads being switched into and out of circuit, changes in the output voltage can cause very serious problems.

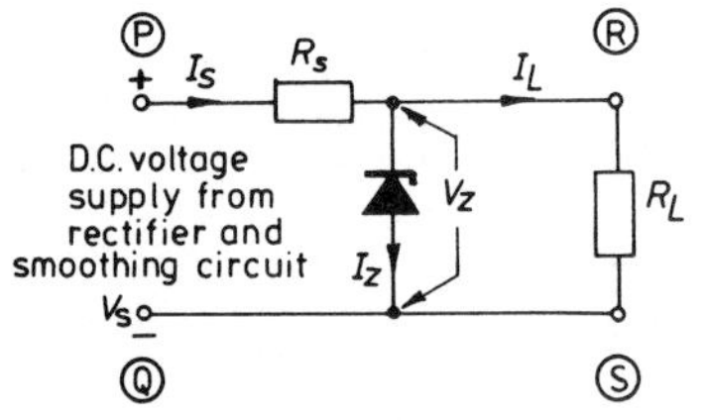

Figure 3.14

The circuit of a simple Zener stabiliser is shown in *Figure 3.14*. This is connected to the output of a power supply unit at terminals P and Q. Resistor R_s is connected in series with the Zener and provides the necessary protection against excessive current flow at the breakdown point. The resistor R_L represents, as usual, the load connected across the diode; this could be a small amplifier, for example, or perhaps an electric motor for a model. Notice how the diode is connected: its cathode goes to the positive terminal of the d.c. supply, hence it is reverse biased and under proper application will be operating in the breakdown condition.

To understand the action of the circuit, the Zener diode can be considered as a reservoir of current so long as it remains broken down. It then responds to variations in both the supply voltage at terminals P and Q, and the load current I_L flowing through R_L as follows:

1. Suppose the d.c. output voltage at P-Q increases for some reason, then the current through the Zener increases while the increase in voltage appears across R_s – *not* across the Zener. The Zener voltage, remember, remains at its breakdown value V_z, irrespective of the increase in current through it. Similarly, if the d.c. output at P-Q decreases, the Zener surrenders the extra current and the voltage across R_s falls. So the variation is 'absorbed' by series resistor R_s and the output voltage at R-S remains constant.

2. If the load current I_L increases for any reason, the Zener current decreases by the same amount. Similarly, if the load current decreases, the Zener current increases by the same amount. This time, the Zener takes up any excess current and sheds any current difference demanded by the load, so acting as a current reservoir while maintaining a constant voltage at terminals R-S.

There is a minimum Zener current for which the voltage stabilisation is effective and the Zener current must never be permitted to fall below this. In other words, the Zener must be in its breakdown condition at all times. This minimum current is a function of the voltage across the Zener and so this condition represents the minimum value of the applied d.c. voltage permissible. Zeners can stabilise satisfactorily down to currents of the order of 0.5 mA or so. The upper limit of current is, of course, dependent upon the power rating of the device.

The following examples are worked for you and illustrate the action of stabiliser circuits in actual figures. You need nothing more than Ohm's law as your mathematical equipment to be able to cope with problems on simple Zener stabilisers.

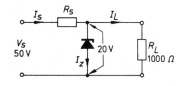

Figure 3.15

Example (4). A 20 V stabilised supply is required from a 50 V d.c. input. A 20 V Zener diode is to be used, having a power rating of 2 W. Find the required value of the series resistor R_s.

The basic circuit of the stabiliser is shown in *Figure 3.15*. Now we have to protect the Zener by making sure that the greatest current flowing through it at its rated breakdown voltage of 20 V does not exceed its power rating of 2 W. The greatest current will flow through the Zener when the load current I_L is zero (or if the load is disconnected). Hence

$$\text{Maximum current} = \frac{\text{Watts}}{\text{Voltage}} = \frac{2}{20} = 0.1 \text{ A}$$

To stabilise at 20 V from a 50 V input, the voltage drop across R_s must be 30 V. Hence

$$R_s = \frac{30}{0.1} = 300 \, \Omega$$

Example (5). If the Zener diode of the previous example stabilises down to a current of 1 mA, calculate the greatest and least supply voltage input between which stabilisation is obtained. The load resistance R_L is 1000 Ω.

For the greatest supply voltage (V_s) the Zener will pass its maximum current of 0.1 A. Then, referring to the figure, and working in milliamps

$$I_L = \frac{20}{1000} = 20 \text{ mA}$$

But

$$I_s = I_z + I_L$$
$$\therefore I_s = 100 + 20 = 120 \text{ mA}$$

Hence

$$\text{Maximum } V_s = (120 \times 10^{-3} \times 300) + 20 \text{ V}$$
$$= 56 \text{ V}$$

For the least supply voltage, the Zener must pass 1 mA. Hence

$$I_L = 20 \text{ mA, as before}$$

and

$$I_s = I_z + I_L = 1 + 20 = 21 \text{ mA}$$

Hence

$$\text{Minimum } V_s = (21 \times 10^{-3} \times 300) + 20 \text{ V}$$
$$= 26.3 \text{ V}$$

Notice the wide variation possible in the voltage supply, i.e. 26.3 V to 56 V, over which the output voltage to the load remains constant at 20 V.

Example (6). Suppose the voltage supply V_s in the previous example is 50 V. What will be the minimum possible value to which the load resistance can be reduced if stabilisation is to remain effective?

When R_L is reduced, the load current I_L increases and I_z

correspondingly falls. I_z must not fall below 1 mA, however, so for this condition

$$I_s = \frac{30}{300} = 0.1 \text{ A } (100 \text{ mA})$$

as for Example (4). Then

$$I_L = 100 - 1 = 99 \text{ mA}$$

$$\therefore R_L = \frac{20}{99 \times 10^{-3}} = 202 \text{ } \Omega$$

ZENER PROTECTION

Another application of Zener diodes is shown in *Figure 3.16*, which shows a method of protecting a sensitive meter or similar apparatus against accidental overload. Suppose the meter has a full scale deflection (f.s.d.) current I_m and that at this current the voltage developed across the instrument is V_m. Resistors R_1 and R_2 are then selected in conjunction with the Zener characteristics so that if the applied voltage V exceeds V_m, the excess current resulting will be bypassed through the diode and not through the meter. This can be arranged by ensuring that when V equals V_m the Zener just breaks down but draws only a negligible current. At this time the meter must, of course, just indicate its f.s.d.

Always work problems of the kind we are encountering in this Unit section by the direct application of Ohm's law and common sense.

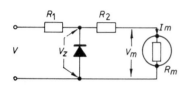

Figure 3.16

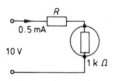

Figure 3.17(a)

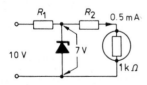

Figure 3.17(b)

Example (7). A meter has a f.s.d. of 0.5 mA and a resistance of 1000 Ω. It is used as a voltmeter by the addition of a series resistor and is scaled 0–10 V. Design a suitable protection circuit using a 7 V Zener diode so that the instrument will not be overloaded by connection to voltage sources greater than 10 V.

There are two parts to this problem: we first have to find the value of the series resistor which converts the 0.5 mA f.s.d. meter into a 0–10 V voltmeter. Look at *Figure 3.17(a)*; when 10 V is applied to the circuit, 0.5 mA must flow through the meter. So

$$R + 1000 = \frac{10}{0.5 \times 10^{-3}} = 20\,000 \text{ } \Omega$$

$$\therefore R = 19\,000 \text{ } \Omega$$

Figure 3.17(b) shows the protection circuit, and here $R_1 + R_2$ must be equal to 19 000 Ω. When the applied voltage is 10 V, the Zener must just break down, but the current through it will be negligible. At this time the meter must indicate f.s.d. The current through R_1, R_2 and the meter in series is therefore 0.5 mA. The situation, in other words, is identical with the circuit at (*a*). Then, for 7 V across the Zener

$$R_1 = \frac{10 - 7}{0.5 \times 10^{-3}} = 6000 \text{ } \Omega$$

and so

$$R_2 = 19\,000 - 6000 = 13\,000 \text{ } \Omega$$

THE VARACTOR DIODE

An increasingly common application of semiconductor diodes is as variable capacitors. Such *varactors* depend for their action upon the depletion layer which forms at the *p-n* junction when zero or reverse bias is applied. As we have already noted, the depletion layer, being an intrinsic region free from charge carriers, acts as an insulator interposed between the *p-* and *n*-type regions. The width of the layer is a function of the applied reverse voltage. Under these conditions the diode acts as a parallel-plate capacitor, the junction contact area and the thickness of the dielectric (the depletion layer) determining the actual capacitance obtained. By variation of the reverse bias, the effective thickness of the dielectric is changed and so is the capacitance. *Figure 3.18* illustrates the principle.

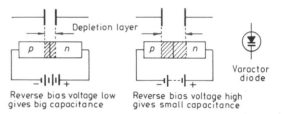

Figure 3.18

Typical capacitance variation is from 50 pF at 1 V reverse bias to 10 pF at 30 V reverse bias. Although the range of capacitance achieved may seem relatively small, these devices have many applications, particularly as tuning elements in such units as television tuners, and as control capacitors in automatic frequency control circuits. Their advantage is their small physical size together with the absence of any mechanical moving parts, the control of capacity being purely the control of a d.c. voltage applied as reverse bias.

PROBLEMS FOR SECTION 3

(8) Complete the following:
 (a) An ideal diode has ...... forward resistance and ...... reverse resistance.
 (b) The power dissipated in an ideal diode would be ......
 (c) A thermionic diode should be operated in the .... limited region only.
 (d) In the circuit of *Figure 3.4* the average output voltage is ...... V, where V is the r.m.s. secondary voltage.
 (e) The addition of a reservoir capacitor to the output of a rectifier circuit ...... the average output voltage and .... the ripple
 (f) In a two-diode full-wave rectifier circuit, the peak inverse voltage is .... the peak input voltage.

(9) A transformer feeding a bridge rectifier has a secondary voltage of 20 V r.m.s. What is the peak inverse voltage across the diodes?

(10) To avoid excessive bit wear when not in immediate use, 240 V mains voltage is applied to a 60 W soldering iron by way of a diode rectifier. Calculate, for this condition: (a) the r.m.s. current flowing, (b) the power dissipated in the iron, (c) the peak current.

(11) A rectifier diode has a constant forward resistance of 5 Ω and a reverse resistance which may be considered infinite. For a 100 V peak input and a 200 Ω load resistor, calculate: (a) the average current flowing, (b) the average voltage across the load, (c) the average voltage across the diode.

(12) What will be the peak voltage across the diode of Problem (11) (a) when it is conducting, (b) when it is not conducting?

(13) Refer to the half-wave rectifier circuit of *Figure 3.4*. What would be the effect on the output waveform shown if a resistor of value equal to R_L was connected (a) in series with the diode, (b) in parallel with the diode?

(14) A sinusoidal voltage $v = 50 \sin 314.2t$ is applied to a circuit made up of an ideal rectifier in series with a 20 Ω resistor. Sketch the current waveform in the resistor over a complete input cycle, carefully indicating the scale values. What is (a) the r.m.s. value of the supply voltage, (b) the r.m.s. value of the rectified current?

(15) The anode of a thermionic diode is at a voltage of 250 V d.c. and the current flowing through the valve is 20 mA. What power is being dissipated at the anode of the valve?

In the following five problems, refer to the circuit of *Figure 3.14*.

(16) Show that the power dissipated in the Zener diode is $(I_s - I_L) . I_L R_L$.

(17) A 20 V stabilised supply is required from a 40 V d.c. input voltage V_s. A 20 V Zener diode is used having a power rating of 1 W. What value resistor is required for R_s?

(18) A Zener diode is to provide a 16 V stabilised output from a 20 V supply. The load resistor $R_L = 200$ Ω and the Zener current $I_z = 8$ mA. Find the value of the series resistor R_s and the power dissipated in it.

(19) If the Zener of Problem (17) stabilises down to a current of 0.5 mA calculate the greatest and least supply voltage V_s between which stabilisation will be obtained. Take $R_L = 2000$ Ω.

(20) A 68 V, 3 W Zener diode is to be used to supply a variable load from a 100 V d.c. source. Find the value of the required series resistor if the Zener is not to be overloaded under any load conditions.

(21) A varactor diode has a relationship between applied voltage V and capacitance C (pF) given by $C = k/\sqrt{V}$ where k is a constant. If $C = 20$ pF when $V = 15$ V, what will be the capacitance when the voltage changes to 10 V?

4 The bipolar transistor

Aims: At the end of this Unit section you should be able to:
Understand the general structure of the bipolar transistor.
Identify the electrodes of a transistor as emitter, base and collector, and know the three modes of connection.
Define the current gains of a transistor connected in common-base and common-emitter mode, and state the relationship between these gains.
Sketch the static characteristics of common-base and common-emitter connected transistors.
Use characteristic curves for the evaluation of circuit impedances and gain.

The *bipolar transistor* is a semiconductor device which can act as an amplifier as well as a switch, which was the basic property of a diode. Because of its amplifying property it is also suited to perform as an oscillator. The transistor can therefore take over all the operations and applications which up to a few years ago were the sole province of the thermionic valve. It differs from the valve, however, in several important respects. In the valve, electrons are emitted from a heated cathode surface situated in an evacuated envelope, and their attraction to a nearby positively charged anode electrode constitutes the flow of current through the device. Control of this internal flow of electrons by means of a suitable electrode interposed between the cathode and anode surfaces leads to the property of voltage amplification. In the transistor, which is our main concern here, electrons (or holes) are injected into the solid material of the semiconductor and their subsequent movement through the body of the material constitutes the flow of current which can be similarly controlled. Operating voltages are considerably lower than those necessary for thermionic valves. Like the semiconductor diode, therefore, most transistors are small and compact elements, many times smaller than even the most miniature of valves.

THE JUNCTION TRANSISTOR

Consider a junction transistor as two junction diodes connected back to back, as illustrated in *Figure 4.1*. From either of these arrangements there are three external connections and these are known as the *emitter*, *base* and *collector* terminals of the transistor. Batteries are connected to the coupled diodes in such a way that the emitter-base diode is biased in the forward (low resistance) direction, and the base-collector diode is biased in the reverse (high resistance) direction. As the diagrams are drawn, diodes D_1 are conducting and diodes D_2 are switched off, in cases *(a)* and *(b)*. In both circuits there will be a small leakage current due to minority carriers moving across the reverse-biased base-collector diode junctions.

Now we shall not obtain the circuit properties of a transistor from two discrete diodes actually connected together in this way. In *Figure 4.1(a)* we have the arrangement *n-p-p-n*, and in *(b)* the arrangement *p-n-n-p* of the semiconductor material types. In both cases the centre portion is made up of two similar type materials, either two *p*-type

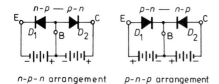

Figure 4.1(a)   (b)

32 *The bipolar transistor*

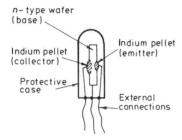

Figure 4.2

anodes as at *(a)* or two *n*-type cathodes as at *(b)*. For transistor action, this centre base region must be a common interface between the outer emitter and collector electrodes. This is accomplished in manufacture by sandwiching a thin base wafer between two electrodes of opposite type material.

Figure 4.2 shows a cross section of a *p-n-p diffused junction transistor* made on the so-called alloying principle. During manufacture, the germanium or silicon is grown into a single crystal and doped with a suitable impurity, such as antimony, to make it *n*-type. It is then sawn and lapped to form a basic wafer whose thickness is of the order of 0.02 mm or less. Both surfaces of this wafer are then converted, to a carefully controlled depth, to *p*-type material by diffusing into them small pellets of indium placed on the surfaces. At a temperature of 155 °C the indium pellets melt and at 550 °C dissolve into the wafer to form a liquid alloy. On cooling, the alloy re-crystallises, leaving a small but sufficient amount of *p*-type impurity in the wafer material. Connecting leads are attached to the two indium pellets and to the wafer between them; the whole assembly, after suitable cleaning and etching treatment, is mounted in a small hermetically sealed container.

There are a great number of other manufacturing processes available which result in electrodes of different sizes and dispositions, but the basic action of the transistor is the same in each case; only such things as power handling capability and high frequency characteristics are affected by these different techniques of manufacture.

The *p-n-p* transistor, like the *n-p-n* transistor, is therefore a sandwich containing a *p-n* and an *n-p* junction connected in series, so that there is a common electrode of either *n*-type or *p*-type material (the base) at the centre of the sandwich. A pictorial representation and the circuit symbols for both types of transistor are shown in *Figure 4.3*, where the arrowhead on the emitter indicates in both cases the conventional direction of current flow in that part of the circuit. Conventional flow inside the transistor is therefore equivalent to the *direction of hole movement*.

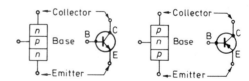

Figure 4.3

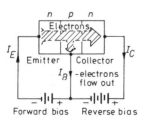

Figure 4.4

In *Figure 4.4* we have the representation of an *n-p-n* transistor with the appropriate forward and reverse biasing voltages applied to the respective parts of the circuits. The application of a negative potential to the emitter causes the electrons in the *n*-type material there to be repelled from the emitter region. The emitter therefore acts as a source of electrons and these flow into the base, which is biased positively with respect to the emitter. We have, in other words, a forward-biased diode. In the base region the electrons drift towards the collector by a process of diffusion and are accepted by the collector which is biased positively with respect to the base. Because the base is *p*-type, electrons exist there only as minority carriers, so the electrons arriving at the collector are derived almost entirely from the emitter supply. On their way across the base wafer, a small proportion (about 1–2%) of the electrons recombine with holes in the base region and this loss of charge is made good by a flow of base current. It is to reduce this 'loss' of electrons that the base is made very thin.

The effect seen from the external circuit is that of a fairly large flow of current (electrons) from emitter across the base to the collector, with a small flow of current from the base. For a small transistor, we might find such typical values as 1 mA for the emitter current I_E, 0.98 mA for the collector current I_C and the difference 0.02 mA base current I_B.

This is the mechanism of an *n-p-n* transistor. For a *p-n-p* device the majority carriers in the *p*-type emitter are holes. The emitter acts as a source of holes which flow into the base when the base is biased negatively with respect to the emitter. The collector is now biased negatively with respect to the base and so absorbs holes from the base region. Here a small proportion of the holes leaving the emitter recombine with electrons which are the majority carriers in the base, and this loss of charge is made good by a flow of electrons into the base as base current. (See *Figure 4.5.*) Compare the *p-n-p* figure with the *n-p-n* figure, noticing carefully the direction of movement of the electron or hole carriers inside the transistor and the corresponding movement of *electron carriers only* in the external circuit. You will probably find it easier to comprehend the *n-p-n* example at first, since the carriers in all parts of the circuit are electrons. Keep in mind that the *p-n-p* circuit simply reverses both battery connections and the external current directions. We will, in general, refer to *n-p-n* type transistors in the notes which follow.

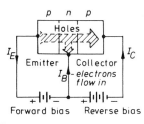

Forward bias   Reverse bias

The indicated current direction is the true electronic (electron) flow. Conventional flow is in the opposite direction

Figure 4.5

Since the carriers originate at the emitter and distribute themselves between base and collector, the sum of the base and collector currents must always be equal to emitter current, so

$$I_E = I_C + I_B$$

CIRCUIT CONFIGURATIONS

There are three possible ways in which a transistor can be connected into a circuit, shown in *Figure 4.6* at (*a*), (*b*) and (*c*). In all cases one electrode is common to what we have called the input and output terminals, and the circuit configuration is described in terms of this common electrode. Hence (*a*) shows the *common-base* connection, (*b*) the *common-emitter* connection, and (*c*) the *common-collector* connection. The common electrode is usually treated as being at earth potential and the term 'earthed' or 'grounded' instead of 'common' is sometimes used to define the particular configuration; for example, (*a*) might be referred to as the 'grounded-base' connection.

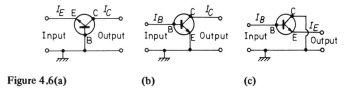

Figure 4.6(a)   (b)   (c)

Current Gain

Let us consider the relationship existing in each configuration between the input and output currents, assuming that we have battery supplies appropriately connected to the various terminals. We may define the ratio

$$\frac{\text{Current flowing in the output circuit}}{\text{Current flowing in the input circuit}}$$

as the static current gain of the transistor, the word static being used to imply that we are operating our devices solely from direct durrent sources, i.e. batteries, no other external components being involved.

Current gain in the common-base configuration is designated a_B* and is the ratio of collector current I_C to emitter current I_E. So

$$a_B = \frac{\text{Collector current}}{\text{Emitter current}} = \frac{I_C}{I_E}$$

In common-emitter configuration, current gain is designated a_E and is the ratio of collector current I_C to base current I_B. So this time

$$a_E = \frac{\text{Collector current}}{\text{Base current}} = \frac{I_C}{I_B}$$

There is a relationship between a_B and a_E because for all configurations the equation $I_E = I_C + I_B$ must be true. From the above definitions

$$\frac{a_B}{a_E} = \frac{I_B}{I_E} = \frac{I_E - I_C}{I_E}$$

since $I_B = I_E - I_C$. Hence

$$\frac{a_B}{a_E} = 1 - \frac{I_C}{I_E} = 1 - a_B$$

So, rearranging:

$$a_E = \frac{a_B}{1 - a_B}$$

By transposition

$$a_B = \frac{a_E}{1 + a_E}$$

You should remember these two important basic relationships. Now try the following problems.

(1) An arrow on the emitter symbol that points towards the base indicates a (an) .... transistor.
(2) The circuit configuration in which the input is between base and emitter and the output is between collector and emitter is the .... connection.
(3) An n-p-n transistor requires a .... collector operating voltage.

*The symbols h_{FE} and h_{FB} are commonly used for static current gain, but we shall retain the alternative symbols a_E and a_B in this book.

> (4) A transistor has an emitter current of 2.5 mA and a collector current of 2.4 mA. What is its base current?
>
> (5) For the transistor of the previous problem, calculate its current gain for (i) common-base, (ii) common-emitter connection.
>
> (6) What current ratio defines the gain of the common-collector configuration? (Look at *Figure 4.6(c)* if you wish.) Prove that this current gain, a_C, is equivalent to $1 + a_E$.

You should have discovered one or two important facts from working these test problems, in conjunction with what has gone before. For one thing, the ratio

$$a_B = \frac{I_C}{I_E} \text{ is always less than } 1$$

for the reason that I_C is always less than I_E. In the same way, the ratio

$$a_E = \frac{I_C}{I_B} \text{ is always greater than } 1$$

for the reason that I_C is always greater than I_B.

Further, the current gain a_C of the common-collector configuration is clearly always greater than 1, since $a_C = 1 + a_E$.

Typical figures for the current gains of common-base and common-emitter connections are 0.97 to 0.998 and 30 to 500 respectively.

> (7) A transistor has a common-emitter current gain of 100. What will be its gain in common-base?
>
> (8) The common-base gain of a transistor is 0.985. What is its gain in common-emitter?

From the point of view of current gain, the common-base connection seems of no value as an amplifier, but it has other advantages which make it a relatively common configuration in practical electronic circuits.

CHARACTERISTIC CURVES Because the transistor is a device having three terminals, there are a number of measurements which may be made on it in comparison with the simple relationship between current and voltage which is made on the semiconductor diode.

In the notes which follow the conventional procedure for identification of the various circuit voltages and currents will be followed. It is customary to fix the common rail (in most cases this is the earth or chassis line) at 0 V and to measure all voltages relative to this. So for an *n-p-n* transistor, voltages measured will be positive, and for a *p-n-p* transistor they will be negative. The exception is the voltage measured between emitter and base in common-base configuration when the base is connected directly to the zero rail. Voltages will be identified

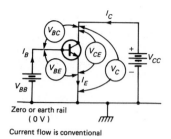

Figure 4.7

by two subscripts if they act between two definite terminals of the transistor, or if they are voltage sources such as battery supplies; and with a single subscript when they act relative to the zero rail. *Figure 4.7* shows this convention in use. V_C, for example, refers to the collector voltage relative to the zero rail, but V_{CE} refers to the voltage acting between collector and emitter. If the emitter is connected directly to the zero rail as shown in the figure, clearly V_C would be identical to V_{CE}, but would differ if, for instance, a resistor was connected in either the emitter or collector lead. Capital letters are used throughout and indicate that d.c. measurements are being considered.

There are four quantities, or *parameters* as they are called, which can be either held at a constant value or varied during the course of a measurement experiment, and these are: (a) input voltage, (b) input current, (c) output voltage and (d) output current. Four characteristic graphs may be drawn involving pair combinations of these so as to describe the d.c. performance of the transistor completely.

These *static characteristics* are usually found for particular transistors in the manufacturer's literature. The three most important characteristic curves for these two modes of connection will now be discussed.

The Input Characteristic The input characteristic of a transistor is a plot of input current against input voltage, hence it concerns only the base-emitter junction which is the equivalent of a forward biased diode. For the common-base configuration, where the 'live' input terminal is the emitter and the base is at the zero rail potential, the emitter current I_E is measured for a range of values of base-emitter voltage V_{BE}, the collector being maintained throughout the measurement at a constant voltage V_{CB}.

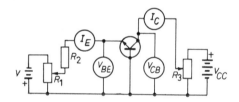

Figure 4.8(a)

Figure 4.8(a) shows a suitable circuit arrangement for such an experiment. V_{CB} is fixed by a suitable setting of R_3, usually to a value at which the transistor would be used in practice, and V_{BE} is varied in suitable steps by potentiometer R_1. R_2 is simply a safety resistor included to prevent excessive base current being drawn. Corresponding values of I_E are then recorded from the milliammeter wired in series with the emitter input. A graph of I_E against V_{BE} can then be plotted in the manner shown in *Figure 4.8(b)**. Notice that the value assigned to V_{CB} during the measurements is stated on the graph paper, in our case by way of example $V_{CB} = 6$ V.

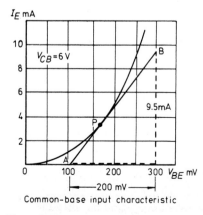

Common-base input characteristic

Figure 4.8(b)

*In all graphs, voltage and current axes will always be indicated positive, so that the first quadrant of the co-ordinate axes will contain the required curves. This is quite conventional, although strictly when negative values are involved, other quadrants should be used to give a true graphical representation. Some older books show such reversed axes. The shape of the curves and the deductions drawn from them are not affected in any way by the positive axes convention.

For the common-emitter configuration, the 'live' input terminal is the base and the emitter is at the zero rail potential. Hence base current I_B is measured for a range of values of base-emitter voltage V_{BE}, the collector again being maintained at a constant voltage V_{CE}. A suitable circuit is illustrated in *Figure 4.9(a)*, and a typical graph resulting from such an experiment at (b).

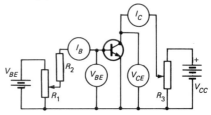

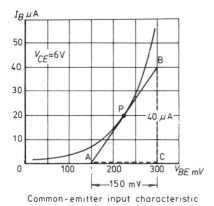

Common-emitter input characteristic

Figure 4.9(b)

Figure 4.9(a)

What information do the input characteristics of transistors give us? One of the chief interests is the *input resistance* of the transistor — what magnitude of resistance would I 'see' if I looked into the input terminals of the transistor? This parameter is very important in the study and design of transistor amplifiers. One thing we can be fairly certain about — the input resistance is likely to be relatively low because we are looking into a forward biased diode. However, by the proper use of the input characteristic we can find out quite accurately what the actual working resistances for the two modes of connection will be.

Return to *Figure 4.8(b)*, the common-base input curve. The graph gives us related values of I_E and V_{BE}; we could therefore select any point on the curve, read off the pair of related values represented by the point, and by the division V_{BE}/I_E obtain the usual Ohm's law value of resistance at that point. However, the graph is markedly non-linear and we should clearly obtain a different value for the resistance at every point selected. We are more concerned with the ratio

$$\frac{\text{The small change in } V_{BE}}{\text{The small change in } I_E}$$

because when the transistor is used as an amplifier, the input signal is not a d.c. level but a small alternating quantity superimposed on a fixed d.c. level. The choice of the d.c. level depends upon the exact circuit requirements and once selected it is referred to as the d.c. operating point P. The above ratio is then found as the gradient of the curve at the point P. The value for input resistance we then obtain is called the *dynamic input resistance* of the transistor under actual signal conditions, unlike the static resistance found as the simple division V_{BE}/I_E. The dynamic input resistance is then

$$R_i = \frac{\text{The small change in } V_{BE}}{\text{The small change in } I_E} = \frac{\delta V_{BE}}{\delta I_E}*$$

for a constant V_{CB}.

*This is the shorthand mathematical symbolism for the ratio. The δ simply means 'a small change in ...' For those who are not familiar with this symbolism and would like a further note on the subject of curve gradients, there is an Appendix on page 146 and it is strongly suggested that you read this before going any further in the text.

In *Figure 4.8(b)*, suppose the operating point P is at an emitter current of about 3.5 mA. Through this point a tangent AB is drawn; *the gradient of this tangent is the same as the gradient of the curve in the immediate neighbourhood of point P.* Now the ratio previously given is the ratio of the horizontal change in V_{BE}, the length AC, to the corresponding change in I_E, the length BC. Hence the changes considered are 200 mV in voltage for 9.5 mA in current. Therefore

$$R_i = \frac{AC}{BC} = \frac{200 \times 10^{-3}}{9.5 \times 10^{-3}}$$

$$= \frac{200}{9.5} = 21\,\Omega$$

In *Figure 4.9(b)* we have the input characteristic in common-emitter mode. Since the vertical scale is now in microamperes, it is obvious that the input resistance will be much higher than in common-base mode. Taking a point P corresponding to a base current of 20 µA and drawing the tangent AB as before, it is seen that

$$R_i = \frac{AC}{BC} = \frac{150 \times 10^{-3}}{40 \times 10^{-6}}$$

$$= \frac{150 \times 10^3}{40} = 3750\,\Omega$$

The values obtained for R_i for the common-base and the common-emitter modes are quite typical; actual values of course depend upon the transistor, the working voltages and selection of the operating point.

> (9) Referring to *Figures 4.8(b)* and *4.9(b)*, estimate the dynamic input resistance for each mode of connection, assuming that the operating point P is at I_E = 2 mA for common-base, and at I_B = 10 µA for common-emitter. (Use a pencil to avoid spoiling your book diagrams.)

The Output Characteristic

The output characteristic is possibly the most useful of transistor characteristic curves because it not only enables us to find the *output resistance R_o* of the configuration but several other useful parameters may be derived from it.

The output characteristic is a plot of collector current I_C against collector voltage for fixed values of base current in common-emitter mode or emitter current in common-base mode. It is usual to draw a family of curves for each configuration over a range of fixed values for I_E or I_B. The circuit of *Figure 4.8(a)* can be used now to plot the output characteristic of the common-base mode. I_E is fixed at, say, 2 mA by suitable adjustment of potentiometer R_1, and V_{CB} is then varied in a series of voltage steps by R_3. Corresponding values of I_C are then recorded for each step in V_{CB} from the milliammeter wired in series with the collector lead. A graph of I_C against V_{CB} can then be plotted as shown in the lower curve of *Figure 4.10*. The experiment is repeated for other fixed values of I_E; in the figure these values are

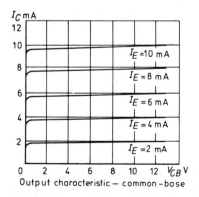

Output characteristic — common-base

Figure 4.10

taken to be in 2 mA intervals up to a maximum of 10 mA. In this way a family of curves is obtained which may, if necessary, be interpolated for estimation of collector voltage and current at other values of emitter current. Each curve is basically the same shape and the portions above a fraction of a volt of applied V_{CB} are practically horizontal. Notice that for each setting of I_E the collector current is almost identical to I_E, the small difference in each case being the base current.

> (10) Referring to *Figure 4.10*, can you explain why collector current flows even when V_{CB} is zero?

For the common-emitter configuration, the circuit of *Figure 4.9(a)* may be used again. Since the base-collector junction is a reverse biased diode, a small leakage current will flow when $I_B = 0$. The lowest curve in *Figure 4.11* shows this effect; if the applied V_{CE} was increased sufficiently, the junction would break down at the avalanche (Zener) point. Now if by adjustment of potentiometer R_1 a small base current is set to flow, say, of 20 μA, and held constant while V_{CE} is adjusted throughout the available range of voltage, the next characteristic curve of the family is obtained. It is basically the same shape as before, reasonably flat, but not so flat as was the characteristic of the common-base mode, and the collector current is at a much higher level. For further settings of I_B, further curves are obtained. Again, when presented with such a family of output chracteristics we may interpolate to estimate collector current and voltage at other base current values from those shown.

Both sets of curves immediately tell us that the collector current which flows is substantially independent of collector voltage — the curves are practically horizontal lines. This is because collector current is derived from the charge carriers originating in the emitter. On entering the base under the emitter-base voltage these carriers are almost entirely gathered up by the collector, even if the collector-base voltage is very small. Increases in the collector voltage, therefore, have little effect on collector current because it is not the collector voltage which is producing the carriers. The horizontal portions of the curves represent *transistor saturation regions*. As far as the output terminals are concerned, the transistor behaves as a *constant current generator*, and this property has several useful applications in electronic circuit design.

Return now to *Figure 4.11*. The curves give us related values of I_C and V_{CE} for various values of I_B. The output resistance of the transistor is the effective resistance we should see by looking back into the output terminals; we should expect this to be relatively large because this time we are looking into a reverse biased diode. However, the value we measure is not simply the straightforward resistance of the collector-base diode junction. Working on the same principle as we did for the case of input resistance, we can say

$$R_o = \frac{\text{The small change in } V_{CE}}{\text{The small change in } I_C} = \frac{\delta V_{CE}}{\delta I_C}$$

and a triangle ABC drawn on the appropriate characteristic enables R_o to be evaluated. Strictly, the hypotenuse of the triangle, AB, is a tangent to the curve, but as the lines are for all practical purposes straight and

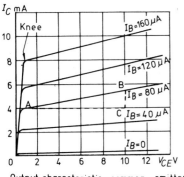

Output characteristic – common – emitter

Figure 4.11

parallel, the characteristic can be considered as the actual hypotenuse. What we are saying is that the gradient of the lines is constant. From the diagram then

$$R_o = \frac{AC}{BC} = \frac{9}{1.5 \times 10^{-3}} = 6000 \text{ } \Omega \text{ } (6 \text{ k}\Omega)$$

Clearly, the output resistance increases as the characteristic lines become more horizontal, so referring to the curves for the common-base mode shown in *Figure 4.10*, it is apparent that the output resistance for this mode of connection will be very high. A typical value would be 500 kΩ. The common-emitter, on the other hand, would have a typical output resistance of 20 kΩ; the curves of *Figure 4.11* were deliberately drawn to illustrate the calculation for R_o, and the value of 6 kΩ derived above is, in general terms, rather low.

The Transfer Characteristic The transfer characteristic is a plot of the relationship between output current I_C and input current I_E or I_B, depending upon the mode of connection, at a specified collector voltage. The characteristic can be plotted directly from related values measured from the experimental circuits of *Figures 4.8* or *4.9*, or may be graphically derived from the respective output characteristics already discussed.

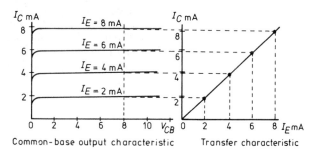

Figure 4.12

The output characteristics of *Figure 4.10* can be used to determine the common-base transfer characteristic in the manner shown in *Figure 4.12*. For a specified collector-base voltage V_{CB}, in this example taken to be 8 V, values of I_E are projected across to the co-ordinate axes shown on the right. Notice that the vertical axis of I_C is common to both graphs. Where these horizontal projections meet the verticals for corresponding values of I_E, points lying on the transfer characteristic are established. Joining these intersections gives us the required characteristic.

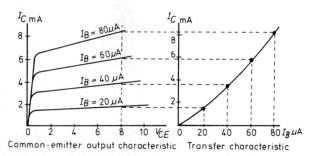

Figure 4.13

The common-emitter transfer characteristic can be obtained similarly from an output characteristic of the form of *Figure 4.11*, as shown in *Figure 4.13*. Here the specified collector-emitter voltage V_{CE} is again taken as 8 V. Values of I_B are projected across to the co-ordinate axes on the right of the figure, the vertical axis of I_C again being common to both graphs. Where these horizontal projections meet the verticals for corresponding values of I_B, points lying on the transfer characteristic are established, and the line joining them can be drawn in.

From the graphs, the gradients of the transfer characteristics give us

$$\frac{\delta I_C}{\delta I_E} = a_B; \qquad \frac{\delta I_C}{\delta I_B} = a_E$$

the current gains in common-base and common-emitter mode respectively.

You will notice that the common-base transfer characteristic is much more linear than is the common-emitter. Since I_E is always $I_C + I_B$ and I_C is plotted against I_E in *Figure 4.12*, the line will have a gradient which is only slightly less than 45°; a_B in other words is always slightly less than 1, since the tangent of 45° is 1.

> (11) Use the transfer characteristic of *Figure 4.13* to obtain a rough estimation of the common-emitter current gain a_E.

Now work through the following test problems, making sure you are acquainted with the d.c. operating conditions of transistors in common-base and common-emitter modes of connection before going on to your study of transistor amplifiers in the next Unit section.

PROBLEMS FOR SECTION 4

(12) Complete the following statements:
  (a) In normal operation the emitter-base diode is ...... biased and the collector-base diode is ...... biased.
  (b) The carriers crossing the base of an *n-p-n* transistor are ....
  (c) A *p-n-p* transistor requires a .... collector operating voltage.
  (d) The arrow on the emitter of an *n-p-n* transistor points ...... the base.
  (e) The current gain of the .... mode is always less than unity.
  (f) The input resistance of the common-base connection is much .... than is the input resistance of the common-emitter connection.
  (g) $I_C + I_B$ is always equal to ....
(13) What are the values for a_E transistors having the following values of a_B (a) 0.980, (b) 0.975, (c) 0.970, (d) 0.96?
(14) Calculate a_B for transistors having the following values of a_E: (a) 50, (b) 75, (c) 110, (d) 350.
(15) A transistor has an emitter current of 1.5 mA and a base current of 10 μA. What will be its a_B and a_E values?

(16) Are the following statements true r

(16) Are the following statements true or false?
(a) In common-base mode: (i) I_E is dependent upon I_C; (ii) V_{EB} is dependent upon V_{CB}; (iii) I_C is independent of V_{CB}.
(b) In common-emitter mode, no collector current flows if the base is disconnected.
(c) Carriers injected by the emitter into the base region are of the same polarity as the collector.
(d) Collector voltage influences the number of carriers injected into the base by the emitter.

(17) If $I_C = 4.0$ mA when $V_{CE} = 2.0$ V, and 5.0 mA when 9.0 V, I_B being held constant, calculate the value of the output resistance under this condition.

(18) If $I_C = 5.0$ mA when $V_{CB} = 8.0$ V and 5.02 mA when 12 V, I_E being held constant, estimate a value for the output resistance of the transistor.

(19) In a common-emitter connected transistor, with V_{CE} held constant, a change of 80 mV in V_{BE} caused a change of 65 μA in I_B. What is the input resistance of the transistor under this condition?

(20) In a common-base transistor circuit, with V_{CB} held constant, V_{EB} is 120 mV when I_E is 1.0 mA, and 200 mV when 7.0 mA. What is the input resistance of the transistor?

(21) The table gives the input characteristic parameters of a small transistor in common-emitter mode, for $V_{CE} = 4.5$ V.

V_{BE}	75	100	125	150	175	200 (mV)
I_B	3	8	20	40	72	110 (μA)

Plot the input characteristic carefully, and from it evaluate the input resistance of the transistor when $I_B = 40$ μA.

(22) Describe an experiment to determine the input current/voltage, and the output current/voltage characteristics of a transistor connected in common-emitter configuration. Explain how the input and output resistances of the transistor can be deduced from these curves.

(23) The data given in the table refer to a transistor connected in common-emitter mode:

Collector volts V_{CE}	Collector current, mA			
	Base current, 20 μA	Base current, 40 μA	Base current, 60 μA	Base current, 80 μA
3 V	0.91	1.60	2.30	3.0
5 V	0.93	1.70	2.50	3.25
7 V	0.97	1.85	2.70	3.55
9 V	0.99	2.04	3.0	4.05

Plot the output characteristics for base currents of 20, 40, 60 and 80 μA and use the curves to determine (a) the current gain a_E when the collector voltage is 6 V, (b) the output resistance when $I_B = 60$ μA.

5 The bipolar transistor as amplifier

Aims: At the end of this Unit section you should be able to:
Understand the operation of a small signal common-emitter amplifier.
Construct the load-line and determine the current and voltage gain from the static characteristic curves.
Explain the necessity for base bias and its stabilisation.
Understand the significance of leakage current and describe thermal runaway.
Calculate the voltage, current and power gains of the amplifier.
Use the bipolar transistor as a switch.

THE GENERAL AMPLIFIER

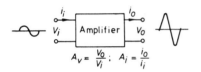

Figure 5.1

An amplifier is essentially a box having two input terminals and two output terminals, as shown in *Figure 5.1*. The box will contain one or more amplifying devices, transistors or valves, together with some associated components such as resistors and capacitors, and some kind of power supply derived from batteries or rectifier units of the kind described earlier.

We expect the amplifier to fulfil two conditions:

1. The output signal will be greater in amplitude than the input signal.
2. The output signal will be of exactly the same waveform (shape) as the input signal.

The first of these conditions is a measure of the voltage or current *amplification* or *gain* provided by the amplifier. We define

$$\text{Voltage amplification } A_v = \frac{\text{Output signal voltage } v_o}{\text{Input signal voltage } v_i}$$

and

$$\text{Current amplification } A_i = \frac{\text{Output signal current } i_o}{\text{Input signal current } i_i}$$

The signal voltages and currents will normally be measured in r.m.s. values, but for sinewave signals it is just as convenient in many cases to find the ratio of the input and output peak values; the resulting figures obtained for A_v and A_i are, of course, unaffected by this.

It is often necessary to know the power gain of an amplifier. This can be calculated from the product of voltage and current gains, so

$$A_p = \text{Voltage gain} \times \text{Current gain} = A_v.A_i$$

The second of the conditions means that the signal waveform should not suffer any *distortion* during the process of amplification. It is not a simple matter to design an amplifier having negligible distortion, at least not without sophisticated and expensive circuit systems. For simple amplifiers of the kind we shall be discussing in this section some

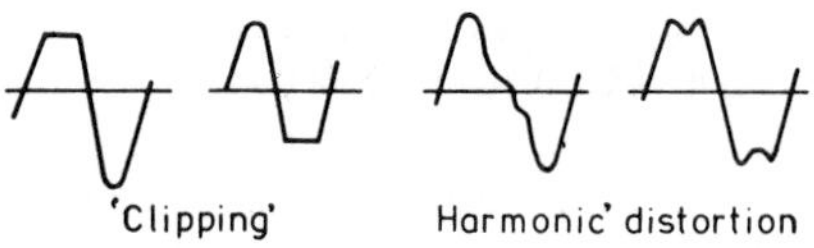

'Clipping'   Harmonic distortion

Figure 5.2

distortion is inevitable, but it can be kept to a reasonably low level by careful attention to certain fundamental details.

Figure 5.2 shows some typical forms of distortion appearing at the output terminals of amplifiers, assuming a pure sinewave input. Some of the reasons for such distortion will become evident as we proceed.

Amplifiers can be classified into two main types:

1. *Small signal amplifiers*, which are designed to amplify small input signals, probably voltage levels of the order of a few microvolts to a few millivolts. An amplifier which immediately follows a record pick-up head or a microphone would be a small signal amplifier, as would an amplifier whose input was the very small radio-frequency signals received on an aerial. It is easier to avoid distortion in amplifiers of this kind than it is in the second category.

2. *Power amplifiers*. These have very large input signal voltages, of the order of several volts. Their output requirements are large current and voltage excursions so that considerable power is available for driving such devices as loudspeakers or, in industrial applications, small electric motors.

At this stage, we are interested only in small signal amplifiers dealing with relatively low frequency signals.

A single amplifier stage (one transistor or one valve) is rarely sufficient to supply the overall amplification needed. It is then necessary to use two or more devices in *cascade*, the output of one being fed into the input terminals of the next, and so on. The signal is then progressively amplified as it passes through the system. At the end, when a sufficient amplification (in the small signal sense) has been achieved, a power amplifier is introduced to provide the required final output level.

In this Unit section we shall investigate the amplifying properties of the transistor. As you have learned, a transistor consists of three layers of *n*- and *p*-type material; these layers form respectively the collector, base and emitter electrodes. The current flowing between the base and the collector can be controlled either by a current flowing in the emitter or in the base circuit, so the transistor is essentially a current operated device. A transistor is not capable on its own of providing amplification, but when it is employed in conjunction with an external *load resistor*, amplification becomes possible. The conventional amplifier arrangement thus consists of a source of power supply (batteries, etc.), a load, and the control device (transistor), connected in series as shown in *Figure 5.3*. The load does not have to be a resistor, it may take the form of a tuned circuit, for example, but for our immediate purposes we will present it as a resistor. From the diagram, the voltage amplification A_v is seen to be

$$A_v = \frac{\text{Output signal voltage across the load}}{\text{Signal voltage present at the input}} = \frac{v_o}{v_i}$$

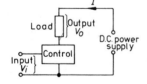

Figure 5.3

THE COMMON-EMITTER AMPLIFIER

For the two basic circuit configurations we discussed in the previous Unit section, we saw that the input resistance in both cases was the relatively low resistance of the forward-biased base-emitter junction; the output resistance was the relatively high resistance of the reverse-biased base-collector junction. It is this marked disparity between the input and output resistances, in conjunction with the external load resistor

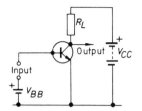

Figure 5.4

connected into the collector (output) circuit, which enables the transistor to act as an amplifier of alternating currents and voltages.

Figure 5.4 shows the basic circuit arrangement of a common-emitter amplifier. In the circuit, input takes place between base and emitter and the output is developed between collector and emitter. The emitter is the common electrode, and it is generally simpler to refer to the connection mode as base-input, collector-output amplification.

In the usual manner, the base is biased positively relative to the emitter (we are using an *n-p-n* transistor) by the battery V_{BB} and the collector is biased positively relative to the emitter (and the base) by battery V_{CC}. A load resistor R_L has been included in the collector lead. All this accords with what we have already discussed about basic transistor operation. The addition of R_L is the only major change in the general circuit arrangement. However, our previous work has dealt with the transistor only from the point of view of the d.c. battery conditions and on this basis (with no load resistor in circuit) we defined the static current gain

$$a_E = \frac{\text{Change in } I_C}{\text{Change in } I_B} \quad \text{with } V_{CE} \text{ constant}$$

On this basis also, we drew the various characteristic curves relating the various electrode voltages and currents.

We are concerned now with what happens when an alternating signal is applied to the input terminals; this is the signal that we require to amplify, be it speech, music or a simple single-frequency tone. For simplicity we shall consider the alternating input to be a single frequency sinewave, expressed in the customary way as $i = \hat{I}.\sin \omega t$. If the emitter-base voltage is allowed to alternate about a mean value, base current I_B also will vary about some mean value determined by battery V_{BB}. It is clear that the emitter current and hence the collector current will also vary about some mean value determined by V_{CC}. The load resistor R_L will have a p.d. developed across it by this alternating collector current, and this will represent the output alternating voltage from the amplifier. The processes involved are perhaps best brought out by way of a simple numerical example.

First, we recapitulate on the d.c. set up, assuming that the a.c. signal input is zero. *Figure 5.5(a)* shows the arrangement. As usual we have the base-emitter junction biased in the forward direction by V_{BB}. This bias is taken, purely as an illustrative example, to be 1 V. This then fixes what we shall call the *base-bias voltage*. Suppose that this bias causes a base current $I_B = 0.1$ mA to flow; then this value of base current determines the mean d.c. level of base current about which the a.c. input signal will swing alternately positive and negative. This is the *base current d.c. operating point*.

Now for this example the static current gain of the transistor, a_E, has been taken to be 50. Since 0.1 mA is the steady base current, the collector current I_C will be given by $I_C = a_E.I_B$, so

$$I_C = 50 \times 0.1 = 5 \text{ mA}$$

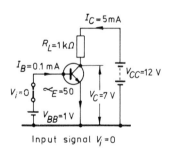

Figure 5.5(a)

This current is shown in the diagram. When 5 mA flows through the load R_L, taken here to be 1 kΩ, there will be a steady voltage drop across R_L given by $I_C.R_L = 5 \times 10^{-3} \times 1000 = 5$ V. The voltage at the collector V_{CE} will therefore be $V_{CC} - I_C R_L = 12 - 5 = 7$ V. This value of V_{CE} is the mean d.c. level about which the output a.c. voltage signal will swing alternately positive and negative. This is the *collector current d.c. operating point*. A transistor biased in this way is said to be operating in *Class A* conditions.

We have now fixed two quantities as operating points: the base current I_B at 0.1 mA, and the collector voltage V_{CE} at 7 V. This establishes the steady or static conditions. We now consider the *working* or *dynamic* conditions.

Turn to (b) of *Figure 5.5* and consider what happens when we apply the a.c. signal input in series with the base. The exact manner of doing this is unimportant at the moment. Suppose the input signal varies between the peak values +0.1 V and −0.1 V as shown, and imagine the condition at (b) to be such that the positive peak of the input is present at the input terminals. Then at this instant V_{BE} is increased by 0.1 V; suppose this causes I_B to increase by 0.05 mA, so that the instantaneous $I_B = 0.1 + 0.05$ mA $= 0.15$ mA. The collector current now increases to $0.15 \times 50 = 7.5$ mA and the voltage drop across R_L increases to $7.5 \times 10^{-3} \times 1000 = 7.5$ V. Hence the collector voltage V_{CE} *falls* to $12 - 7.5 = 5.5$ V. Notice now that for a change of 0.1 V at the base we have produced a change of 2.5 V across the collector load. If we consider the voltage across R_L as the output voltage, then

$$\text{Voltage gain} = \frac{\text{Voltage change across } R_L}{\text{Voltage change in } V_{EB}}$$

$$\therefore A_V = \frac{2.5}{0.1} = 25$$

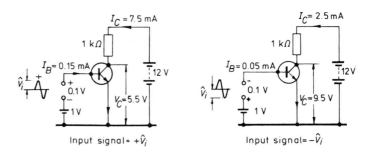

Figure 5.5(b)   (c)

The input signal now changes from its positive peak value through zero to its negative peak value. At the instant it passes through zero, the circuit conditions have returned to those shown in *Figure 5.5(a)* and I_B, I_C and V_{CE} are momentarily at their mean operating values of 0.1 mA, 5 mA and 7 V respectively. In (c) the input signal is at its most negative value, −0.1 V. At this instant V_{BE} is reduced by this amount and I_B decreases by 0.05 mA. The total I_B is now $0.1 - 0.05 = 0.05$ mA. The collector current correspondingly falls to $0.05 \times 50 = 2.5$ mA and the voltage drop across R_L becomes $2.5 \times 10^{-3} \times 1000 = 2.5$ V. Hence the collector voltage V_{CE} rises to $12 - 2.5 = 9.5$ V.

This cyclic variation of V_{CE} about its mean value of 7 V goes on all the time the input signal varies I_B about its mean value of 0.1 mA. The transistor then provides us with a voltage amplification of 25, and the output variation is an exact replica of the input variation, i.e. a sine-wave. Equal changes in I_B have caused equal changes in V_{CE}. There is therefore no distortion occurring in the process of amplification.

You will have noticed that we have interpreted the amplification provided by the transistor in the example as a voltage gain, symbolised A_V. There is also a current gain: I_B is changing by 0.05 mA on each input half-cycle and the output current is changing by 2.5 mA correspondingly, so we have a current gain of 50, or a_E. In an actual amplifier circuit, for reasons which we will not pursue here, the dynamic working gain is always *less* than the static value of a_E, but an appreciable current gain is, nevertheless, provided by the common-emitter amplifier.

> (1) If the load resistor R_L is increased in value, there will be a greater voltage drop across it for a given collector current and hence a greater voltage gain. Is there any reason why R_L should not be made very large, say 1 MΩ, to obtain a large voltage gain?

We should note one other important point at this stage: when V_{BE} increased, V_{CE} decreased, and vice versa. The transistor therefore reverses the phase of the input voltage.

> (2) Does the amplifier reverse the phase of the input current?

USING THE CHARACTERISTIC CURVES

To have a reasonably good idea of the way an actual transistor will work as an amplifier, we have to know a number of the quantities mentioned in the previous illustrative example fairly accurately: the exact point, for example, to which I_B should be set, what amplitude of input signal we expect, what collector supply voltage is available and which transistor we are going to use. All these things are interrelated and the choice of any one of them has to be made with an eye on the others.

Our starting point is the characteristic curves we discussed in the previous Unit section. We are considering the common-emitter amplifier here, so the static characteristics relating to the common-emitter configuration will be our concern.

We look first at the output characteristic curves, a set of which were given earlier in *Figure 4.11* and which are reproduced here as *Figure 5.6*

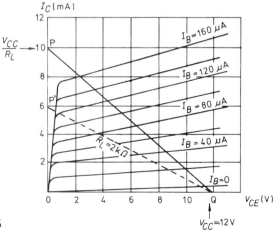

Figure 5.6

for easy reference. These curves relate collector current I_C to collector voltage V_{CE} for various fixed values of I_B. Now we cannot use these individual curves to solve our problems about amplification, because as soon as a sinusoidal signal is applied at the base terminal, I_B varies over a range of values as our earlier example showed, and the individual curves relate to a constant I_B. It is necessary first of all to draw across the characteristic curves another line, known as the *load line*, which will show us the relationship existing between I_C and V_{CE} when I_B is changing. To draw the load line it is necessary for us to know

1. The value of the collector load resistor R_L.
2. The collector supply voltage V_{CC}.

Of course, these values themselves also depend to a certain degree on other factors, but we have to make a start somewhere. Consider the circuit of *Figure 5.7*, where $R_L = 1.2$ kΩ and $V_{CC} = 12$ V. These are typical figures for a small amplifier stage. The collector voltage is given by

$$V_{CE} = V_{CC} - I_C R_L$$

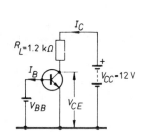

Figure 5.7

Now the supply voltage V_{CC} determines the value of the collector voltage when $I_C = 0$. Clearly, under this condition there is no voltage drop across R_L and so

$$V_{CE} = V_{CC}$$

This establishes the position of point Q on the characteristic curve V_{CE} axis in *Figure 5.6*, and obviously V_{CE} can never exceed this value. Since $V_{CC} = 12$ V, point Q represents this voltage limit.

Now suppose $V_{CE} = 0$. This can only happen if the voltage drop across R_L is exactly equal to V_{CC}. For the resistance given, 1.2 kΩ, and a voltage drop equal to 12 V, the collector current

$$I_C = \frac{V_{CC}}{R_L} = \frac{12}{1200} \text{ A} = 10 \text{ mA}$$

This condition establishes the position of point P on the I_C axis of the characteristic curves. By joining the points P and Q with a straight line we obtain the load line for the condition $R_L = 1.2$ kΩ.

For every given load resistance there will be a corresponding (and different) load line. If V_{CC} is kept at the same value, all the possible lines will start at the same point Q but will cut the I_C axis at different points P. Suppose, for example, R_L is increased to 2 kΩ. When the voltage dropped across this load is 12 V, the current flowing through it will be 12/2000 A = 6 mA, hence the load line for $R_L = 2$ kΩ will lie between Q and point P' as shown in the broken line. It is evident that the gradient of the load line (its steepness) *decreases* as R_L *increases*, and vice versa.

(3) Using the curves of *Figure 5.6*, draw load lines (lightly in pencil) for the following conditions:

(a) $V_{CC} = 10$ V; $R_L = 1$ kΩ, $R_L = 2$ kΩ
(b) $V_{CC} = 8$ V; $R_L = 1$ kΩ, $R_L = 1.5$ kΩ

The bipolar transistor as amplifier 49

> Now check your answers and rub out the lines!
> (4) We have plotted load lines using two extreme points P and Q. How do we know that the lines connecting these are, in fact, *straight* lines?

So far, so good. We have seen how a load line can be drawn to suit a particular value of load resistor. We have now to see how a particular load line will help us in establishing the proper mean values of I_B and V_{CE} about which the input and output signal alternations will respectively swing. We have to find a value of I_B so that, when the signal is superimposed upon it, the current and voltage variations at the collector lie within the limits determined, and imposed, by the extremities of the load line. Quite plainly the collector voltage can never exceed V_{CC} (point Q on the line) and the collector current can never be greater than that value which would make V_{CE} zero (point P on the line).

In *Figure 5.8* we have taken the output characteristics for a transistor amplifier which is to be used with a 1.4 kΩ load and a V_{CC} of 14 V. The

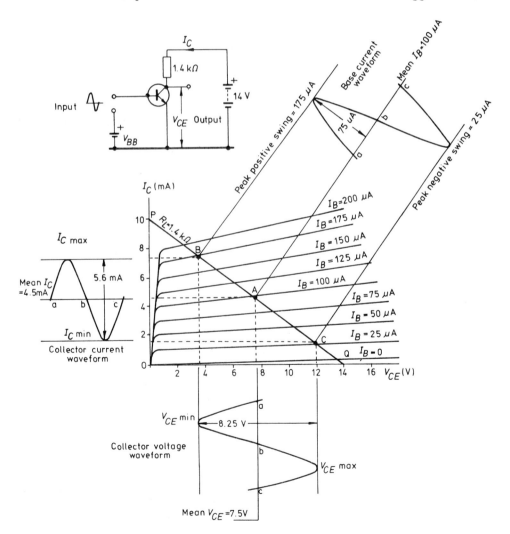

Figure 5.8

load line for $R_L = 1.4$ kΩ has been drawn in between points P and Q. The point A has been chosen as the base current operating point, i.e. $I_B = 100$ μA, because the excursions of I_B about this point which will result in proportional changes in I_C are very closely those between which the load line cuts through equal spacings between the static characteristics on either side of A. The distance from A to B and from A to C, in other words, represents the peak permissible swings of base current under the stipulation that AB is very closely equal to AC. Any appreciable swing beyond these limits will exaggerate the discrepancy between AB and AC, or, worse still, will carry the transistor beyond collector current cut-off at end Q or work into the curved portions of the characteristics at end P. If this is allowed to happen, the output signal will no longer be identical in shape with the base current variations and distortion will be present. It is not usually possible to get A exactly midway between the limits B and C, particularly if a very large input swing is to be handled, so a small amount of distortion is tolerated as the price for a large output signal.

The instantaneous values of the base current waveform are now projected on to the load line to give the corresponding values of I_C; the instantaneous values of I_c (i_c) are likewise projected on to the load line to give the corresponding values of instantaneous output voltage (v_o). Notice that the output voltage waveform is in antiphase with the input current waveform; when base current changes in a positive direction from 100 μA to 175 μA, the collector voltage changes in a negative direction from 7.5 V to about 3.7 V. As base current is in phase with base voltage, the output voltage is antiphase to the input voltage, as we deduced from our earlier example.

From the diagram the maximum excursions of base current for closely equal spacings in either direction about A, where $I_B = 100$ μA, is 75 μA. The corresponding excursions of I_C are about 2.8 mA about the mean value of 4.5 mA. The swing in V_{CE} is then about 4.15 V about the mean value of 7.5 V. These mean values of voltage and current are often referred to as the *quiescent values*, since they represent the no-signal operating conditions. As a rough and ready rule, a mean operating value for V_{CE} is taken to be $V_{CC}/2$ for the purpose of quick calculations.

The signal variations shown in *Figure 5.8* represent the maximum we can allow for this particular circuit if serious distortion is to be avoided. If you measure AB and AC, you will find them approximately equal. Of course, there is nothing to prevent us from making the input signal smaller so that the swing in I_B might be only from 100 μA at A down to 75 μA on one peak and up to 125 μA on the other. The equality of distance along the load line about A would then be almost ideal. What we should get would simply be a smaller output, but with reduced distortion. The operating point must therefore be chosen with care, and once chosen, *stabilised* against any effects which might tend to move it. We shall come to this problem in a very little while.

(5) What is the current gain, A_i, in the amplifier of *Figure 5.8*?
(6) Suppose the input current of this circuit flows in an input resistance of 1000 Ω. What is the approximate voltage gain, A_v?

LEAKAGE CURRENT

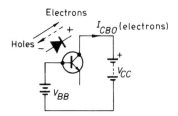

Figure 5.9

Before going any further into general amplifier theory, it is necessary to look into the problem of leakage current. Look at *Figure 5.9*, which shows an *n-p-n* transistor in common-base configuration but with its emitter lead disconnected. Under this condition it might appear that the collector current I_C would be zero, since clearly I_E is zero, but this is not so. The collector circuit is still connected through the reverse-biased base-collector diode, and this diode must pass reverse current. This is where some of the facts you learned in Unit section 2 should stand you in good stead. The leakage current is due to the movement of minority carriers (holes in the *n-p-n* transistor) across the junction, and its direction is opposite to that of the main forward current which would flow if the diode was forward-biased. But a movement of holes from collector to base inside the transistor is equivalent to a movement of electrons in the direction base-to-collector, hence in the external circuit the leakage current flows in the *same sense as that due to collected electrons*, the majority carriers. This leakage current is denoted by I_{CBO}, meaning that we are referring to the collector-base junction with O showing that the third electrode, the emitter, is left disconnected. This current still flows when the emitter is reconnected and the main forward current from the emitter is superimposed. So far we have taken the collector current to be $I_C = a_B I_E$. But with the addition of the leakage current, the true total collector current becomes $I_C = a_B I_E + I_{CBO}$, as *Figure 5.9* shows.

As I_{CBO} is very small at room temperatures (20–25 °C), particularly with silicon material, it might seem that the small addition to the normal relatively large forward current at the collector would be unimportant. This is true for the situation just described, but leakage current increases with increasing temperature and when we look at the problem in relation to the common-emitter configuration, such an increase leads to very undesirable effects in the transistor performance.

We illustrate the common-emitter situation in *Figure 5.10*. Here the base current is treated as the input. Since $I_E = I_C + I_B$ we can write the previous expression for I_C as

$$I_C = a_B(I_C + I_B) + I_{CBO}$$

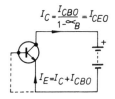

Figure 5.10

Then

$$I_C(1 - a_B) = a_B I_B + I_{CBO}$$

$$\therefore \quad I_C = \frac{a_B}{1 - a_B} I_B + \frac{1}{1 - a_B} \cdot I_{CBO}$$

Now

$$a_E = \frac{a_B}{1 - a_B}$$

$$\therefore \quad I_C = a_E I_B + \frac{I_{CBO}}{1 - a_B}$$

The first term here is the value of I_C we have so far taken as the output of the common-emitter amplifier with the input current equal to I_B. Refer back to page 34 if your memory has faded at this point. The second term must represent the *leakage current when the base is disconnected*; this is denoted I_{CEO}. Hence

$$I_{CEO} = I_{CBO} \times \frac{1}{1-a_B}$$

Hence, for common-emitter connection

$$\text{Total } I_C = a_E I_B + I_{CEO}$$

This is identical in form to the expression for I_C in common-base mode, but the values of the parts are quite different. Suppose, for example, that $a_B = 0.98$, so that $1/(1-a_B) = 50$. Then I_{CEO} is 50 times as large as I_{CBO}, which clearly aggravates the problem we mentioned earlier about the increase in leakage current with rise in temperature. What is happening, in fact, is that in common-emitter mode, the transistor is amplifying its own leakage current! This is definitely a most undesirable state of affairs and a matter of serious concern in transistor applications. Try the next two problems to see if your ideas are straight about this leakage problem.

> (7) If $I_C = 2.45$ mA, $I_{CBO} = 20$ µA and $I_E = 2.5$ mA, what is a_B?
>
> (8) If I_B is zero for the above transistor, what is I_C?

SETTING THE OPERATING POINT

We return now to the topics we were discussing in relation to operation along the load line PQ of *Figure 5.8*. Having selected the operating bias point A on a particular load line drawn on a particular set of characteristics, it is necessary to set the bias current of the transistor concerned to the required value and ensure that it stays there. In *Figure 5.8* and in the earlier diagrams we assumed that a separate bias battery V_{BB} was used for this purpose, but in practical designs it is not convenient to have this arrangement and in general the base bias is obtained from the same source as the collector supply, that is, a single battery (V_{CC}) provides all the necessary voltages throughout the amplifier. As the required bias voltage is much smaller than that required at the collector, the most simple modification is shown in *Figure 5.11*. Here the forward base bias is obtained by the insertion of a resistor R_B between the base and the positive terminal of the supply. The value of this resistor is easily calculated: in our example from *Figure 5.8*, the required I_B is 100 µA, hence

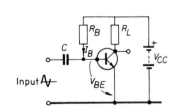

Figure 5.11

$$R_B = \frac{V_{CC}}{I_B} = \frac{14}{100 \times 10^{-6}} = 140 \text{ k}\Omega$$

This value actually includes the internal base-to-emitter resistance, but as this is of the order of some few hundred ohms, it may be ignored in the calculation. R_B consequently limits the base current in the forward direction to the required value, in this case 100 µA. It is important to notice, however, that the bias is not developed across R_B but across the base-emitter junction as the result of the no-signal (d.c.) current through that junction. This action makes the base positive with respect to the emitter, so biasing the diode in the forward direction.

(9) A transistor requires an operating base bias of 50 μA. The V_{CC} supply is 9 V. What value of resistor R_B is required to provide this bias?

(10) The above transistor has $a_E = 70$, and is used with a collector load R_L of 2000 Ω. What will be the steady voltage at the collector?

Now the use of a single biasing resistor like this is not particularly good practice from the point of view of maintaining stability in collector current. Suppose the temperature increases, then I_{CBO} also increases and I_{CEO} becomes a_E times this variation. Hence I_C increases, and V_{CE} and I_B are also influenced; for the d.c. collector current

$$I_C = a_E I_B + I_{CEO}$$

and the d.c. base current $= I_B - I_{CEO}$.

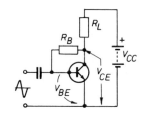

Figure 5.12

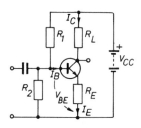

Figure 5.13

So, with any increase in leakage current, the total collector current increases and the total base current decreases. The operating point is consequently unstable, and the additional heating at the collector might lead to the effect of *thermal runaway* or at least to an increase in distortion. We shall discuss this in detail in a later note.

An alternative method of base biasing is shown in *Figure 5.12*. Here the resistor R_B is returned, not to the V_{CC} line, but to the collector itself. If now the collector current increases for any reason, the collector voltage V_{CE} will fall. As the base bias resistor is taken from the collector the base current will also fall, since $I_B = V_{CE}/R_B$. Hence the collector current $I_C = a_E I_B$ will also fall and tend to restore itself to its original (pre-rise) value.

Figure 5.13 shows the most commonly used bias arrangement. It consists of a potential divider circuit made up from resistors R_1 and R_2 connected in series across the supply, and an emitter resistor R_E. If the potential divider is made up so that the voltage level at the centre point (the base connection) is that required to establish the proper base current, but at the same time the total value of $R_1 + R_2$ is such that the current flowing through the divider is very large compared with I_B, then the base current will remain substantially constant regardless of variations in collector current. The emitter resistor in turn determines the value of emitter current which will flow for a given base voltage at the junction of R_1 and R_2. Any increase in I_C will produce an increase in I_E and this in turn will increase the voltage drop across R_E. This reduces the forward bias voltage V_{BE} which then leads to a reduction in I_C, so partly compensating for the original increase. This argument applies equally well to changes in collector current resulting either from changes in a_E (which happen when a transistor is changed) or in the supply voltage. Thus this circuit gives better d.c. operating stability than one in which the emitter is connected directly to the earth line. It is usual to make R_E of such a value that a drop of about 0.5–1 V occurs across it, and to proportion the divider so that R_2 is about 5 to 10 times the value of R_E. The total current through the divider should normally be at least ten times the mean value of base current.

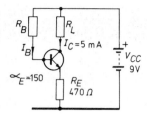

Figure 5.14

Example (11). A transistor with $a_E = 150$ is used in the circuit of *Figure 5.14.* The bias conditions are such that a collector current of 5 mA flows in the collector load. Ignoring leakage current and taking $V_{BE} = 0.65$ V, calculate a suitable value for R_B.

$$\text{Emitter current } I_E = \frac{I_C}{a_B} = \frac{a_E + 1}{a_E} \cdot I_C$$

$$= \frac{151}{150} \times 5 \times 10^{-3} = 5.033 \text{ mA}$$

$$\text{Emitter voltage} = I_E R_E = 5.033 \times 10^{-3} \times 470$$

$$= 2.366 \text{ V}$$

$$\therefore \text{ Base voltage} = 2.366 + 0.65$$

$$= 3.02 \text{ V}$$

$$\therefore \text{ Required voltage drop across } R_B = 9 - 3.02 = 5.98 \text{ V}$$

$$\text{Now} \quad \text{Base current } I_B = \frac{I_C}{a_E} = \frac{5 \times 10^{-3}}{150} = 33.3 \text{ }\mu\text{A}$$

$$\therefore R_B = \frac{5.98}{33.3 \times 10^{-6}} = 180 \text{ k}\Omega$$

Example (12). Figure 5.15 shows a common-emitter amplifier with potential divider bias and an emitter resistor. The quiescent base current is 50 µA and the base-emitter voltage is 0.6 V. If the voltage drop across R_E is to be 1 V, assess suitable values for R_1, R_2, R_E and R_L.

For a 1 V drop across R_E at the indicated value of $I_E = 1$ mA $R_E = 1$ kΩ. The current through the divider has to be large relative to I_B, so taking the current in R_2 to be $10 I_B$ or 0.5 mA, the p.d. across $R_2 = V_{BE} + I_E R_E = 0.6 + 1.0 = 1.6$ V.

$$\therefore R_2 = \frac{1.6}{0.5 \times 10^{-3}} = 3200 \text{ }\Omega$$

$$\text{Current in } R_1 = 11 I_B = 0.55 \text{ mA}$$

$$\therefore \text{ p.d. across } R_1 = V_{CC} - 1.6 = 10.4 \text{ V}$$

$$\therefore R_1 = \frac{10.4}{0.55 \times 10^{-3}} = 19 \text{ k}\Omega$$

As there is a 1 V drop across R_E, 11 V is available for the drop across R_L and V_{CE} in series. It is reasonable to take the mean V_{CE} at about the midpoint of this supply, hence $V_{CE} = 5.5$ V. So, taking $I_E = I_C = 1$ mA

$$R_L = \frac{5.5}{10^{-3}} = 5500 \text{ }\Omega$$

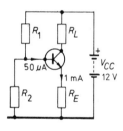

Figure 5.15

THE TRANSISTOR AS A SWITCH

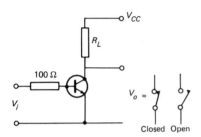

Figure 5.16

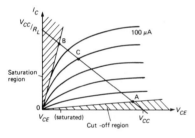

Figure 5.17

We have seen earlier how a diode behaves as an automatic switch, OPEN to forward voltages and CLOSED to reverse voltages. A transistor can also be used as a switch because it can be changed from a high resistance in its cut-off state (switch open) to a low resistance in its saturated or 'bottomed' state (switch closed). Let us see how these two conditions come about.

A transistor is switched on by injecting current into its base. This current is amplified by the transistor and the collector current increases as the base current increases, finally reaching a limiting value determined by the transistor itself and the associated components. The base-emitter diode of the transistor, when conducting, has a voltage drop across it of about 0.6 V (for silicon) and this is substantially constant irrespective of the current passing. By placing current-limiting resistors in both the base and collector leads, a practical transistor switch becomes as shown in *Figure 5.16*. When V_i is at zero volts, no base current flows and the transistor is non-conducting. The output voltage V_o is then at V_{CC} level. If now V_i rises to 1 V, say, the base-emitter voltage is 0.6 V and the remainder of the applied input (0.4 V) develops across the 100 Ω base resistor. The current flowing into the base is therefore 0.4/100 A or 4.0 mA. This switches the transistor fully on and the collector voltage falls to a very low value, typically 0.1 V.

Both on and off conditions are best illustrated by a look at the output characteristic curves shown, for a typical transistor, in *Figure 5.17*. When the input voltage is zero, the operating point is located at point A; the cut-off current region is exaggerated for clarity. The switch, whose 'contacts' are the collector and emitter terminals, is then OPEN. When the input voltage is positive, the operating point moves to point B. Any further increase in base current above (in this example) 100 μA produces no further increase in collector current; it remains at a value given approximately by V_{CC}/R_L. This is the 'saturated' state. The switch 'contacts' are then closed. Note that at a point on the load line such as C, the transistor is switched on but *not* saturated. All such intermediate positions on the line do not produce an effective switch which has a low contact resistance.

The applications of the transistor as a switch are many, particularly in logic where circuits operate strictly under ON or OFF conditions.

MUTUAL CONDUCTANCE

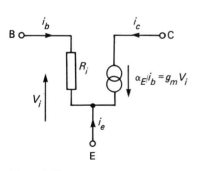

Figure 5.18

The effect of a variation in base current (which occurs when a signal is applied to the base) appears in the collector circuit as a magnified (amplified) version of the input, the amplification being a_E. We can imagine this collector current being produced by a hypothetical current generator located within the transistor, so we can draw a model or *equivalent circuit* to represent this (*Figure 5.18*). Such models will form a large part of your later work in the analysis of amplifier systems. The input current i_b (small letters are used here to indicate that this current is changing) flows into the base-emitter diode of the transistor which has an input resistance R_i. The collector current is this input current multiplied by a_E so that $i_c = a_E i_b$, and we imagine this current to be internally generated by the source shown.

Instead of working in terms of the input current, however, we might work in terms of the input voltage v_i acting between base and emitter. This is a particularly useful approach as it is easier to measure the input voltage for experimental purposes, and the transistor is often used as a voltage amplifier anyway. From the diagram, $v_i = R_i i_b$ or $i_b = v_i/R_i$.

$$a_E i_b = a_e \frac{v_i}{R_i}$$

and

$$i_c = \frac{a_E}{R_i} v_i$$

This equation shows us that the collector current i_c can be calculated from a knowledge of the base-emitter *voltage* input. For this purpose we introduce a parameter called the *mutual conductance* (or transconductance) of the transistor, symbolised g_m. There is a similar parameter for the unipolar (field effect) transistor to be discussed later. Mutual conductance is the ratio of the change in collector current to the change in the base-emitter voltage which causes it. Since g_m is the ratio current/voltage, its dimensions are those of Siemens, or amps/volt.

It can be proved that g_m for a transistor is given closely by $I_C/25$ millisiemens (where I_C is in milliamps). This equation is perhaps more conveniently expressed as

$$g_m = 40 I_C \text{ mA/V or mS}$$

Thus for a collector current of 1 mA (standing bias value here), g_m = 40 mA/V, that is, a 1 V change at the base will produce a 40 mA change in the collector current.

This parameter is particularly useful because it does not depend upon the transistor but only upon the collector current setting.

THERMAL RUNAWAY

Thermal runaway is an effect which arises from the presence of leakage current in a transistor and can lead rapidly to the destruction of the transistor unless steps are taken to prevent its occurrence. Because of the amplifying effect on the leakage current in the common-emitter circuit, this mode of connection is much more susceptible to thermal runaway than is the common-base mode. We can best illustrate the effect by means of a closed loop diagram, *Figure 5.19*. Suppose there is an increase in the temperature of the base-collector junction — someone may have left his transistor radio receiver in bright sunlight or on top of a heated radiator! The increase in temperature shown at A then leads to an increase in leakage current as at B. This leakage current in turn leads to an increase in the total I_C, as at C, and this in itself raises the temperature of the junction by permitting more power to be dissipated at the collector, as at D. This brings us back to our starting point, so the initial temperature rise is augmented by the following train of events. This is known as a case of *positive feedback*; the events in the closed loop become self-sustaining and unless some action is taken to 'break' the chain at some point, the transistor current rises to such a high value that the base-collector diode avalanches out of control and the transistor is rapidly destroyed.

Apart from ensuring that the stability of the operating point is good, thermal runaway can be prevented by mounting the transistor on some form of heat sink which will conduct heat away from the junction as fast as it is produced. The important factor is the maximum junction temperature permissible before thermal agitation within the semiconductor material sets off the rise in leakage current which leads to thermal runaway. This maximum is about 85 °C for germanium and 150 °C for silicon, so germanium is most susceptible to temperature variations.

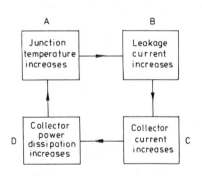

Figure 5.19

For small transistors the junction is not normally in close thermal contact with the case. A simple form of heat sink can be used in such cases, taking the form of a clip-on corrugated aluminium tube or a copper 'flag' (*Figure 5.20*). These devices aid the loss of heat from the transistor case by increasing the contact area with the surrounding air.

Push-on corrugated heat sink   'Flag' heat sink   Power transistor

Figure 5.20

Power transistors are usually made so that the junction is in very close thermal contact with the case, the case itself being designed so that it can be bolted directly to a large metal surface. We have already mentioned this kind of cooling in connection with power rectifier diodes on page 21. *Figure 5.20* also shows a typical power transistor of this sort and the thick copper or aluminium base construction which forms an integral part of an external heat sink.

GAIN ESTIMATIONS

Although voltage and current gains can be estimated from the characteristic curves, it is sometimes sufficient to obtain values from simple formulae. The results are, of course, only approximations, but they are adequate for most general purposes. They are based on a knowledge of R_L, R_i and a_E. Input resistance R_i can be deduced from the input characteristic for a given V_{CE} as explained on page 37.

Assume a collector signal current of i_c so that the signal voltage across R_L is $i_c R_L$. If the current gain is a_E, the base signal current $i_b = i_c/a_E$. Now for an input resistance R_i, the input voltage which must be applied across R_i to produce a base current of i_c/a_E is $i_c R_L/a_E$. Hence the voltage gain is given by the ratio

$$A_v = \frac{i_c R_L}{i_b R_i} = a_E \frac{R_L}{R_i}$$

The power gain of the amplifier can be evaluated similarly. The input signal power is given by $P_i = i_b^2 . R_i$ and the power delivered to the collector load by the transistor is $P_o = i_c^2 R_L$. Hence the power gain

$$A_p = \frac{P_o}{P_i} = \frac{i_c^2 . R_L}{i_b^2 . R_i} = a_E^2 \frac{R_L}{R_i}$$

$$= a_E \frac{v_o}{v_i}$$

A summary of the various gains for the common-emitter amplifier (and the common-base) follows, together with typical values of input and output resistance. You might care to try to verify some of these results for the common-base amplifier on the lines given for the common-emitter in this Unit section. You will find suitable common-base characteristics in the previous section.

58 The bipolar transistor as amplifier

	Common-emitter	Common-base
Input resistance, R_i	1 kΩ–3 kΩ	10 kΩ–500 kΩ
Output resistance, R_o	10 kΩ–50 kΩ	100 kΩ–1 MΩ
Current gain, A_i	High	Unity
Voltage gain, A_v	High	High
Power gain, A_p	High	Medium
Phase shift	180°	Zero

PROBLEMS FOR SECTION 5

All problems will relate to n-p-n common-emitter amplifiers, unless otherwise indicated or implied.

(13) Are the following statements true or false?
 (a) Power gain may be expressed as A_v^2/R_L.
 (b) If the junction temperature increases, V_{BE}, I_{CBO} and a_E all increase.
 (c) If two transistor amplifiers each having a voltage gain of 10 are connected in cascade, the overall voltage gain is 20.
 (d) I_{CBO} approximately doubles for every 10°C rise in temperature.
 (e) For a common-emitter amplifier, the change in I_C due to a change in I_B is $a_E \times$ change in I_B.
 (f) If the load resistor is continually reduced in value, the gain of the amplifier continually approaches the value given by a_E.

(14) Plot the following characteristic curves:

V_{CE}	I_C (mA)			
	I_B = 20 μA	40 μA	60 μA	80 μA
3	0.91	1.60	2.30	3.00
5	0.93	1.70	2.50	3.25
7	0.97	1.85	2.70	3.55
9	1.00	2.05	3.00	4.05

The transistor concerned is used in a common-emitter amplifier with R_L = 2500 Ω and V_{CC} = 10 V. Draw the load line for these conditions. From your diagram estimate (a) the value of I_B for V_{CE} = 5 V; (b) the current gain when V_{CE} = 5.5 V; the output voltage swing if I_B swings between 20 μA and 80 μA.

(15) A transistor is connected in common-emitter mode with R_L = 12.5 kΩ. The static current gain a_E is 100 and the input resistance is 1000 Ω. Find the voltage and power gains of the transistor.

(16) The data given in the following table refer to a transistor in common-emitter configuration. Plot the output characteristic curves.

V_{CE}	I_C (mA)		
	$I_B = 40\ \mu A$	$80\ \mu A$	$120\ \mu A$
2	2.2	4.4	6.4
4	2.4	4.8	7.0
6	2.6	5.2	7.6
8	2.8	5.6	8.2
10	3.0	6.0	8.8

Estimate (a) the output resistance of the transistor for $I_B = 80\ \mu A$; (b) the collector current when $I_B = 100\ \mu A$ and $V_{CE} = 6$ V.

(17) The transistor of the previous problem is used as an amplifier with $V_{CC} = 12$ V and is biased to a base current of $80\ \mu A$. If the steady collector voltage V_{CE} is 6 V under this condition, estimate from the curves (a) the collector current, (b) the value of load resistor R_L.

(18) Refer to the curves you drew for Problem (14). Using the same load line conditions as before, determine the voltage gain, current gain and power gain when an input current of $30\ \mu A$ peak varies sinusoidally about a mean value of $50\ \mu A$. You can take the input resistance as 1 kΩ.

(19) A circuit of the form shown in *Figure 5.14* has $a_E = 120$ and $V_{BE} = 0.7$ V. Estimate the values of collector load R_L and bias resistor R_B, if the mean (quiescent) collector current and voltage values are 9 mA and 4.5 V respectively.

(20) A circuit of the form shown in *Figure 5.15* has $V_{CC} = 12$ V, $I_C = 5$ mA, $a_E = 50$, $V_{CE} = 4$ V, $V_{BE} = 0.5$ V, and the voltage drop across R_E is 2 V. Find the values of R_L, R_E, R_1 and R_2. (You may assume that the bleed current through the base potential divider is ten times I_B.)

(21) A common-emitter amplifier has $R_L = 4.7$ kΩ and is supplied from a battery $V_{CC} = 15$ V. If the collector current is 2 mA, what power is dissipated at the collector under quiescent conditions? What power is dissipated in the load resistor under the same conditions?

6 The unipolar transistor

Aims: At the end of this Unit section you should be able to:
Compare the properties of the unipolar transistor with bipolar transistors.
Describe the basic construction of unipolar transistors and explain their principles of operation.
Explain the difference between depletion and enhancement modes.
Determine the form of the output and transfer characteristics and derive the basic operating parameters from these.
Calculate the stage gain of a common-source amplifier with a resistive drain load.

We turn now to the unipolar or field-effect transistor (FET). This is a device which exhibits certain characteristics that are markedly superior to those of bipolar junction transistors and which operates on the principle that the effective cross-sectional area and hence the resistance of a conducting rod of semiconductor material can be controlled by the magnitude of a *voltage* applied at the input terminals.

The FET operates upon a completely different principle from bipolar transistors. In these, the junction has been in series with the main current path from emitter to collector, and the operation of the transistor has depended upon the injection of majority carriers from the emitter into the base region. There is no such injection in FETs which depend only upon one effective *p-n* junction and only one type of charge carrier. For this reason FET's are known as *unipolar* transistors.

As an amplifier the FET has a very high input impedance comparable with that of a thermionic valve, generates less noise than the ordinary transistor, has high power gain and a good high frequency performance. In addition it has a large signal handling capability, voltage swings at the input being measured in volts. At the best, base voltage swings on bipolar transistors is measured in fractions of a volt.

There are several forms of FET and these are discussed below.

THE JUNCTION GATE FET

In its simplest form, the junction gate FET (or JUGFET) is constructed as shown in *Figure 6.1*. Here a length of semiconductor material, which may be either *p*- or *n*-type crystal, has ohmic (non-rectifying) contacts made at each end. The length of semiconductor is known as the *channel* and the end connections form the *source* and the *drain*. We shall assume throughout this discussion that the channel is *n*-type material as this form of construction is the most commonly used in practical designs.

With no voltages applied to the end connections, the resistance of the channel $R = \rho l/A$ where ρ is the resistivity of the material and l and A are the length and cross-sectional area of the channel respectively. For example, if ρ = 5 ohm-cm, l = 0.1 cm and A = 0.001 cm², the

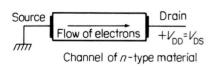

Figure 6.1

channel resistance is 500 Ω. If the source end of the channel is effectively earthed and the drain end is taken to a positive potential, a current will flow along the channel (conventionally) from source to drain. This is drain current I_D. Clearly, if the effective cross-sectional area of the channel can be varied by some means, its resistance and hence the drain current will be brought under external control.

A means of varying the resistance of the channel is shown in *Figure 6.2(a)*. Two *p*-type regions, known as *gates*, are positioned one on each side of the channel. If these two gates are connected to each other and to the source terminal, they form reverse-biased diode *p-n* junctions with the channel crystal and depletion layers will be established as shown in *Figure 6.2(b)*. As the channel has finite conductivity, there will be an approximately linear fall in potential along the channel from the positive drain terminal to the zero (earth) potential at the source terminal, hence the contours of the depletion layers will take the form shown in *Figure 6.2(b)*. As the channel has finite conductivity, there reverse bias voltage between channel and gate is greatest at that end. The flow of electrons from source to drain is now restricted to the wedge-shaped path shown which represents a channel of reduced cross-sectional area compared with the normal condition of diagram (a).

The effective area of the channel is clearly dependent upon the drain-source potential, V_{DS}, because if this potential is increased, the depletion regions will grow and eventually meet; the channel conduction area is then reduced to zero at a point towards the drain end of the channel as *Figure 6.2(c)* depicts. The channel is now said to be *pinched-*

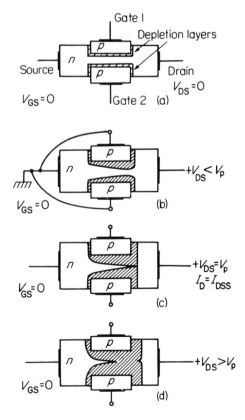

Figure 6.2

off, and the value of V_{DS} at which this occurs is known as the *pinch-off* voltage, V_p.

It is important to take note of the fact that the drain current does not cease when the drain voltage reaches pinch-off because a voltage equal to V_p still exists between the pinch-off point and the source, and the electric field along the channel causes the carriers (electrons) in the channel to drift from source to drain. However, because of the high effective resistance of the channel, the drain current does become substantially independent of the drain voltage.

As V_{DS} is increased beyond V_p the depletion layer thicken between the gate and the drain as shown in *Figure 6.2(d)* and the additional drain voltage is effectively absorbed by the increased field in the wider pinched-off region. The electric field between the original pinch-off point and the source remains substantially unaffected, hence the channel current and so the drain current remains constant. Electrons which arrive at the pinch-off point find themselves faced by a positive potential and are swept through the depletion layer region in exactly the same way as electrons are swept from base to collector in a bipolar n-p-n transistor.

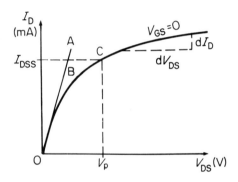

Figure 6.3

It is now possible to represent the relationship between the drain current I_D, and the drain-source voltage V_{DS} in graphical form. In *Figure 6.3* the line OA represents the behaviour of the channel acting simply as a semiconductor resistor; the drain current I_D would follow this line for increases in V_{DS} if the resistance of the channel was constant, unaffected that is, by the effect of the increased drain voltage upon the width of the depletion regions. At point B the action of the depletion layers begins to take effect and there is a departure from the linear characteristic. At point C the applied V_{DS} reaches V_p and pinch-off occurs. For values of V_{DS} greater than V_p the channel remains pinched off and I_D becomes virtually constant and independent of V_{DS}. This characteristic curve is for the particular case when the gate-source potential is zero and is so indicated on the diagram. The drain current which flows when $V_{GS} = 0$ and $V_{DS} = V_p$ is indicated as I_{DSS}, representing the saturated (constant) drain current with the input short-circuited i.e. gates connected to source.

Suppose now that the gates are negatively biased relative to the source instead of being at zero potential. The depletion regions will clearly be thicker for a given value of V_{DS} than they were when there was no such negative bias on the gates. As a result, pinch-off and saturation drain current will occur at lower values of V_{DS}, and when

V_{GS} is sufficiently negative, the electric field between the source and the original pinch-off point will be eliminated and drain current will cease to flow. This situation is illustrated in *Figure 6.4(a)*, and this diagram should be compared to the pinch-off condition of $V_{GS} = 0$ in *Figure 6.2(c)*.

In *n*-channel FETs the gate must be negative with respect to the source to achieve pinch-off. The reverse polarity is necessary for a p-channel device. The sign of V_p is therefore as follows:

n-channel $V_p < 0$
p-channel $V_p > 0$

In operation as an amplifier the FET is biased so that $V_{DS} > V_p$ and $V_{GS} < V_p$. This condition is illustrated in *Figure 6.4(b)*.

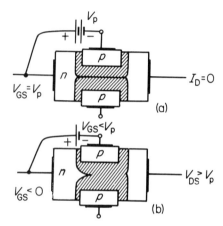

Figure 6.4

Characteristic Curves There are two characteristic curves of interest to use here, the *output* or *drain characteristic* and the *transfer* or *mutual characteristic*.

We have already touched on the form of the output characteristic in discussing *Figure 6.3* and we can now extend that diagram to include the effect of an increasing negative bias on the gate of the transistor. This has been done in *Figure 6.5* and a family of curves relating I_D to V_{DS} for different values of V_{GS} has been obtained. Notice from this diagram that it is possible to operate the junction FET with a small

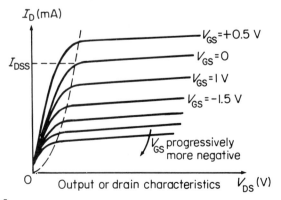

Figure 6.5

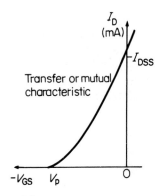

Figure 6.6

positive gate bias. The superficial similarity of these curves to the output characteristics of bipolar transistors should be noted.

The broken line in the diagram represents the locus of a point passing through the respective pinch-off locations on each separate curve. To the left of this line the FET behaves as a variable resistor and this region is known as the *ohmic* region. To the right of the line the FET behaves as a constant current generator and exhibits a very high output resistance, just as the bipolar transistor does. As mentioned, the FET is normally operated in this saturation region where values of V_{DS} exceed V_p.

The transfer or mutual characteristic relates the dependence of drain current I_D upon the gate-source voltage V_{GS} and a typical characteristic is shown in *Figure 6.6*. There is actually a family of these curves, each relating to a particular value of V_{DS}, but as I_D is practically constant beyond the pinch-off point, the curves are almost coincident with one another. To avoid confusion, only a single curve is shown in the figure. Notice that the particular value of I_D for which V_{GS} is zero is I_{DSS}, in accord with the output characteristics of *Figure 6.3* and *Figure 6.4*.

The pinch-off voltage V_p appears on this characteristic as well as on the output characteristics because the channel can be pinched-off by applying V_p between gate and source in the reverse direction. The depletion layers then meet along the whole length of the channel and $I_D = 0$. Strictly, a small leakage current continues to flow along the channel which varies with temperature; this temperature dependence of V_p is due entirely to the variation of barrier potential at the *p-n* junctions, the same effect that causes variation of V_{BE} with temperature in junction transistors. We shall return to problems of temperature effects a little later on.

FET Parameters In order to analyse the amplifying properties of the junction FET, we borrow two parameters from thermionic valve theory. These parameters are mutual conductance and drain slope resistance.

1. Mutual conductance g_{fs} (or g_m) is defined

$$g_{fs} = \frac{\text{the small change in } I_D}{\text{the small change in } V_{GS}} \text{ for a constant } V_{DS}$$

$$= \frac{\delta I_D}{\delta V_{GS}} \bigg|_{V_{DS} = k}$$

This parameter, relating the dependence of drain current I_D to the gate-source voltage V_{GS}, represents the gradient of the transfer characteristic of the FET.

As g_{fs} represents the ratio of a current to a voltage, its unit of measurement will be the Siemen. For a junction FET, g_{fs} will range typically from 10^{-2} to 10^{-3} S.

The equation of the characteristic is complicated, but to a good approximation I_D varies as the square of V_{GS} so that the curve is closely parabolic. It can be shown experimentally that the relationship between I_D and V_{GS} can be expressed in the form

$$I_D \simeq I_{DSS} \left[1 - \frac{V_{GS}}{V_p} \right]^2 \tag{6.1}$$

Notice from this that when V_{GS} is zero, $I_D = I_{DSS}$ as it should.

Example (1) Using the equation just stated, derive an expression for the mutual conductance of a FET. What is the mutual conductance for the case where $V_{GS} = 0$?

Since $g_{fs} = dI_D/dV_{GS}$, the mutual conductance can be obtained by differentiating equation (6.1), noting that V_p is constant. Then

$$g_{fs} = 2I_{DSS}\left[1 - \frac{V_{GS}}{V_p}\right]\left[-\frac{1}{V_p}\right]$$

$$= \frac{-2I_{DSS}}{V_p}\left[1 - \frac{V_{GS}}{V_p}\right] \text{ siemens}$$

The value of g_{fs} obtained at zero bias, or when $V_{GS} = 0$, is denoted by g_{fso} and this follows at once from the previous result:

$$g_{fso} = -\frac{2I_{DSS}}{V_p} \text{ siemens}$$

Notice that g_{fs} or g_{fso} will *always* be positive since either V_p or I will be negative, whether the transistor is *n*- or *p*-channel. The gradient of the mutual characteristic is clearly positive as a glance will show.

(2) Can you see why the expression for g_{fso} above would be useful in a practical method of measuring the g_{fs} of a transistor; *Hint*: substitute g_{fso} back into the equation for g_{fs}, and then think about how you would measure the various parameters.

Example (3). The output characteristics for the 2N5457 FET are shown in *Figure 6.7*. Using these curves, derive the mutual characteristic for this transistor for $V_{DS} = 10$ V. Hence estimate the mutual conductance of the 2N5457 at a gate bias of -0.5 V.

This example will illustrate how the mutual characteristic can be obtained from the output characteristics. From *Figure 6.7* we notice that V_{GS} ranges from zero to -1.5 V while I_D ranges from zero to just over 3 mA. Accordingly we draw the axes for the required mutual characteristic to cover the same ranges. These are shown in *Figure 6.8*.

Referring now to the $V_{DS} = 10$ V point on the output horizontal axis, we make a note of the I_D values where a line erected vertically from this point cuts the V curves; these are the points A, B, C, etc. These I_D values are then plotted on the mutual characteristic axes against the corresponding values of V_{GS}. Connecting the points so obtained, the required characteristic curve appears as the diagram of *Figure 6.8* shows. To find g_{fs} at the point where $V_{GS} = -0.5$ V (point P), we draw a tangent to the curve, and then

$$g_{fs} = \frac{\delta I_D}{\delta V_{GS}} = \frac{2.3 \times 10^{-3}}{1.1} = 2.1 \times 10^{-3} \text{ S}$$

$$= 2.1 \text{ mS}$$

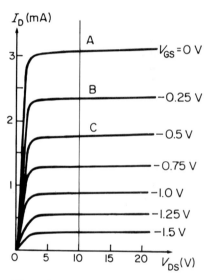

Figure 6.7

66   *The unipolar transistor*

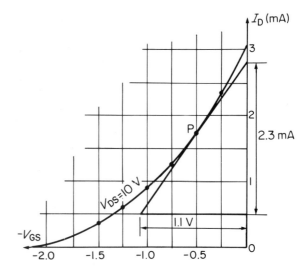

Figure 6.8

As in valve theory, g_{fs} is sometimes expressed in mA per volt; the above solution might equally well be written as 2.1 mA/V. This then tells us that the drain current changes by 2.1 mA when the gate voltage changes by 1 V on a bias of −0.5 V.

We come now to the second of the FET parameters:
2. Drain slope resistance r_d (or r_{ds}) is defined

$$r_d = \frac{\text{the small change in } V_{DS}}{\text{the small change in } I_D} \text{ for a constant } V_{GS}$$

$$= \left. \frac{\delta V_{DS}}{\delta I_D} \right|_{V_{GS} = k}$$

This parameter, relating the dependence of drain current I_D upon the drain-source voltage V_{DS} represents the reciprocal of the gradient of the output characteristic in the saturation region. *Figure 6.3* shows the meaning of drain slope resistance. You should observe that the gradients of the drain characteristics are, like the output curves of a bipolar transistor, very flat in the saturation region and hence the output resistance is very high.

THE JUGFET AS AN AMPLIFIER   Except for the rather different biasing required, the junction FET described above can be used in any of the transistor circuits already discussed. The drain, gate and source are loosely equivalent to the collector, base and emitter of the bipolar transistor, or to the anode, grid and cathode of a thermionic valve. There are, however, two very important differences between the FET and the ordinary transistor which must be emphasised.

First, the control of current flow through the FET is by way of V_{GS}, that is, a *voltage* control and not a current control. Because the input junctions are reverse-biased, the gate leakage current is negligible and the input resistance of the FET is correspondingly very high, of the order of 10^3 to 10^6 megohms. Although the input resistance decreases rapidly when the junctions become forward biased, the input resistance remains relatively high (a megohm or more) in a silicon device provided the forward bias does not exceed some 0.5 V at ordinary temperatures. The FET can therefore be operated as a small signal amplifier with $V_{GS} = 0$.

Secondly, the FET is a majority carrier device; in the *n*-channel form only *electrons* are the carriers drifting from source to drain. In the *p*-channel form only *holes* are the carriers drifting from source to drain, V_{DS} now being of reversed polarity. In both respects, the *n*-channel FET resembles the thermionic valve much more closely than it does a bipolar transistor.

We shall be interested here only in the *common-source* configuration which is the circuit equivalent of the common-emitter connection.

It is useful at this stage to relate the two FET parameters g_{fs} and r_d to a third parameter. We have noted that as far as I_D is concerned, its control can be brought about either by varying V_{DS} (the output characteristic) *or* V_{GS} (the mutual characteristic). The gate voltage V_{GS} exerts a much greater influence on I_D than does the drain voltage V_{DS} for a given variation of potential. Suppose a small increase in V_{DS} causes an increase in I_D, and that I_D can then be restored to its original value by a small negative change in V_{GS}. The ratio of these two changes in the drain and gate voltages which produce *the same change* in I_D is called the *amplification factor* of the FET and is symbolized μ. So

$$\mu = \frac{\text{the small change in } V_{DS}}{\text{the small change in } V_{GS} \text{ producing the same change in } I_D}$$

$$= \left.\frac{\delta V_{DS}}{\delta V_{GS}}\right|_{I_D = k}$$

Since μ is the ratio of two voltages, it is simply expressed as a number. Amplification factor is related to g_{fs} and r_d because the product

$$g_{fs} \times r_d = \frac{\delta I_D}{\delta V_{GS}} \times \frac{\delta V_{DS}}{\delta I_D} = \frac{\delta V_{DS}}{\delta V_{GS}} = \mu$$

Hence

$$\mu = g_{fs} \, r_d \tag{6.2}$$

The circuit, and the circuit symbol, for an *n*-channel FET common-source amplifier is shown in *Figure 6.9*. The input signal V_i is applied between gate and earth, and the drain circuit contains a load resistor R_L across which the output voltage V_o is developed by the flow of drain current. A resistor R_S is included in the source lead and the drain current also flows through this. A voltage equal to $R_S I_D$ is therefore developed across R_S and the source terminal is raised by this potential above the earth line. As the gate is connected to earth through resistor R_G the source is effectively positive with respect to the gate, or, what is the same thing, the gate is *biased negatively* with respect to the source. There is no d.c. voltage developed across R_G because the gate current is negligible; at the same time R_G presents a high impedance

68 The unipolar transistor

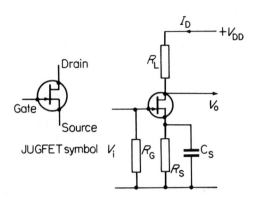

Figure 6.9

to the a.c. input signal. Notice that the drain supply voltage is designated V_{DD}.

Suppose the gate voltage to move positively by a small amount due to the input signal; then I_D will increase and there will be an increased voltage drop across R_L. Hence the drain voltage V_D will fall. Like the bipolar common-emitter amplifier, the common-source FET amplifier introduces a 180° voltage phase change. This negative change in drain voltage in turn causes a further change in drain current, *separate* from but *simultaneous* with the change in drain current due to the original change in gate voltage. The *total* change in I_D is therefore the sum of two simultaneous changes in V_i and V_D. Let a change dV_i in input voltage cause a change dI_D in drain current. Then by definition of g_{fs}

$$\delta I_D = g_{fs}.\delta V_i$$

This change in I_D will in turn produce a change dV_D in drain voltage, so by definition of r_d

$$\delta I_D = \frac{1}{r_d} . \delta V_D$$

But $\delta V_D = -I_D.R_L$, the negative sign indicating that dV_D falls as I_D increases. The total change in I_D is therefore

$$\delta I_D = g_{fs}.\delta V_i - \frac{1}{r_d}.\delta I_D.R_L$$

Rearranging:

$$\delta I_D = \frac{g_{fs}.r_d}{r_d + R_L} \delta V_i = \frac{\mu}{r_d + R_L}.\delta V_i$$

But the output voltage change across R_L is $dV_o = -dI_D.R_L$ hence

$$\delta V_o = -\frac{\mu R}{r_d + R_L}.\delta V_i$$

$$\therefore \frac{\delta V_o}{\delta V_i} = A_v = -\frac{\mu R_L}{r_d + R_L} \tag{6.3}$$

which is an expression for the voltage gain of the FET common-source amplifier.

If we assume that r_d is very much *greater* than R_L (which is true in many amplifiers), the expression for voltage gain approximates to

$$A_v \simeq -\frac{\mu R_L}{r_d} \simeq -g_m . R_L$$

Since the signal input current to a FET is negligible, the current gain is very high but of little importance, and we shall not pursue it further.

Example (4) In a common-source amplifier, the parameters for the FET are $g_{fs} = 2.5$ mS, $r_d = 100$ kΩ. If the load resistor is 22 kΩ, calculate the voltage gain of the amplifier.

From equation (6.3) $A_v = -\dfrac{\mu R_L}{r_d + R_L}$

But $\mu = g_{fs} \times r_d = 2.5 \times 10^{-3} \times 100 \times 10^3$

$$= 250$$

$$\therefore \quad A_v = -\frac{250 \times 2200}{122000}$$

$$= -45$$

It is now possible to derive an equivalent circuit for the FET amplifier, for equation (6.3) connects voltage output V_o, a constant voltage source of e.m.f. $\mu . V_i$ volts, and a resistance made up of r_d and R_L in series. Rewriting the equation gives us

$$V_O = -\mu . V_i \times \frac{R_L}{r_d + R_L}$$

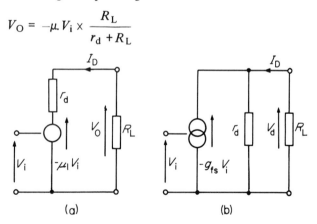

(a)   (b)

Figure 6.10

and this can be represented by the circuit shown in *Figure 6.10(a)*. The FET is replaced by a *constant-voltage* generator whose e.m.f. is $-\mu . V_i$ and this sends current through a resistance made up of r_d and R_L in series, the output voltage V_o being that fraction of $-\mu . V_i$ developed across R_L.

As the FET behaves as a constant-current generator, however, it is often better to use a constant-current form of equivalent circuit and this can be done by expressing equation (6.3) in the form

70 The unipolar transistor

$$V_o = -g_{fs}.V_i \frac{r_d.R_L}{r_d + R_L}$$

where μ has been replaced by the product $r_d.g_{fs}$. This now expresses V_o in terms of a constant current $g_{fs}.V_i$ flowing in a circuit made up of r_d and R_L in parallel. This circuit replaces the FET by a generator which provides a *constant-current* output feeding a parallel arrangement of r_d and R_L. This circuit is shown in *Figure 6.10(b)*. We have assumed for both forms of equivalent circuit that the input resistance of the FET is infinitely high and that the device capacitances are negligible. That is, we are considering a low-frequency equivalent circuit representation.

(5) A voltage gain of 20 is required from a FET having $g_{fs} = 2.0$ mS and $r_d = 50$ kΩ. What should be the value of the drain load resistor to provide this gain?

Example (6) In the amplifier of *Figure 6.9*, $V_p = -2$ V and $I_{DSS} = 2.0$ mA. The circuit is to be biased so that $I_D = 1.2$ mA, the drain supply V_{DD} being 20 V. Estimate (a) V_{GS}; (b) mutual conductance g_{fs}; (c) source bias resistor R_S; (d) the required value of R_L to give a voltage gain of 15.

(a) From our basic equation

$$I_D \simeq I_{DSS}\left[1 - \frac{V_{GS}}{V_p}\right]^2$$

we have by arrangement

$$V_{GS} \simeq V_p\left\{1 - \left[\frac{I_D}{I_{DSS}}\right]^{\frac{1}{2}}\right\}$$

Check this for yourself before going on. Now inserting the given values we obtain

$$V_{GS} \simeq 2\left\{1 - \left[\frac{1.2}{2}\right]^{\frac{1}{2}}\right\} \simeq -0.45 \text{ V}$$

(b) Here we can use the equation

$$g_{fs} = -\frac{2I_{DSS}}{V_p}\left[1 - \frac{V_{GS}}{V_p}\right]$$

and inserting values

$$g_{fs} = -\frac{2 \times 2}{-2}\left[1 - \frac{0.45}{2}\right] = 1.55 \text{ mS}$$

(c) By Ohm's law

$$R_S = \frac{-V_{GS}}{I_D} = \frac{0.45 \times 10^3}{1.2}$$

$$= 375 \text{ }\Omega$$

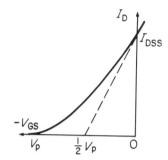

Figure 6.11

(d) We have to use the approximate expression for voltage gain here as we do not know the values of r_d and μ. Assuming that r_d is very large compared with R_L, we have

$$A_v = -g_{fs}.R_L$$

or

$$-15 = -1.55 \times 10^{-3} \times R_L$$

from which

$$R_L = 9.7 \text{ k}\Omega$$

A 10 kΩ resistor would be used here.

(7) *Figure 6.11* shows a mutual characteristic curve for a FET. Prove that if the gradient of the curve at the point where V_{GS} is 0 is continued as the diagram shows, it will cut the horizontal axis at the point $V_p/2$.

(8) The following values are taken from the linear portions of the static characteristics of a FET:

V_{DS}	15	15	10	volts
I_D	13.5	10.5	12.7	mA
V_{GS}	-0.5	-1.0	-0.5	volts

Calculate the parameters r_d, g_{fs} and μ for this FET.

LOAD LINE ANALYSIS

When a load resistor is placed in the drain output circuit to provide a signal voltage output, a load line can be drawn across the output characteristics as it was for the bipolar transistor amplifier. *Figure 6.12* shows a typical graph of this sort, the gradient of the load line depending upon whether an a.c. or a d.c. load condition is being con-

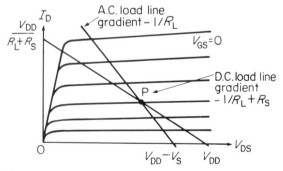

Figure 6.12

sidered. The d.c. load line has a gradient given by $-1/(R_L + R_S)$ and cuts the horizontal axis at V_{DD}, since for $I_D = 0$ the drain voltage must rise to the applied voltage. The other end of the line assumes that the volts drop across the FET is zero at saturation, hence the drain current will be $V_{DD}/(R_L + R_S)$, R_L and R_S being effectively in series across V_{DD}.

The a.c. load line has a gradient given by $-1/R_L$ since R_S is effectively short-circuited by capacitor C_S (see *Figure 6.9*) at signal frequencies, and so cuts the horizontal axis at $(V_{DD} - V_S)$ For Class A operation both load lines must pass through the operating point P which lies on the V_{GS} characteristic corresponding to the d.c. gate-to-source voltage developed across R_S. The procedure for determining the load lines is, in other words, similar to that already described for bipolar transistors and the method of using them to obtain information on the amplifier performance is also identical. The procedure can therefore be summarized here as follows:

(a) The input signal excursions applied to the gate terminal of the amplifier centred on the operating point P must be such that the peaks do not intrude beyond the knee of the $V_{GS} = 0$ characteristic (saturation) or below the horizontal axis limit (cut-off).

(b) The d.c. output drain voltage and drain current at quiescence determine the position of P and this must be chosen with regard to the conditions outlined in (a) above. P will normally be positioned so that the signal swings over characteristic curves which are linear and equally spaced.

THE INSULATED GATE FET

It is the high input resistance of the junction FET which makes it a particularly attractive device in many applications. If an extremely high input resistance is necessary, another type of FET may be employed. This is the metal-oxide semiconductor FET or MOSFET, sometimes also referred to as a MOST or insulated-gate FET, or IGFET.

This device differs from the junction FET in that the gate is actually insulated from the channel by a very thin (about 100 nm) layer of oxide insulation, usually silicon dioxide (glass). The input resistance is then typically in the range 10^6 to 10^8 megohms. MOSFETs are described under two general types: the *depletion* type and the *enhancement* type.

Figure 6.13(a) and *(b)* illustrates the constructional features of the MOSFET. In both forms the gate and channel form the two plates of a capacitor separated by the thin silicon dioxide layer. Because the gate is insulated in this way, V_{GS} can be either positive *or* negative with respect to the channel without conduction taking place through the gate-channel circuit. Any potential applied to the gate establishes a charge on the gate and this induces an equal but opposite charge in the channel.

In *Figure 6.13(a)*, when the gate potential is positive, a negative

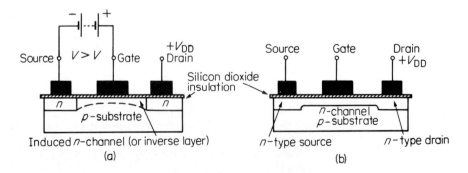

Figure 6.13

charge is induced in the *p*-type substrate at its interface with the silicon dioxide dielectric. This charge repels the majority carriers (holes) from the surface of the substrate and the minority carriers (electrons) that remain form an *n*-channel 'bridge' which connects together the existing *n*-type source and drain electrodes. Consequently, when the drain is connected to a positive supply voltage V_{DD}, electrons flow from source to drain by way of the induced n-channel. Increasing the gate-source voltage V_{GS} *positively* widens or *enhances* the induced channel and the flow of drain current increases. For this reason this form of MOSFET is known as an *n*-channel enhancement FET.

When drain current flows along the channel, a voltage drop is established and this tends to cancel the field set up across the dielectric by the positive gate bias. (Think about the workings of the JUGFET at this point). When the cancellation is sufficient almost to eliminate the induced *n*-channel layer, the channel pinches off and the drain current saturates at a value which is substantially independent of any increases in drain voltage. The output and mutual characteristic curves for an

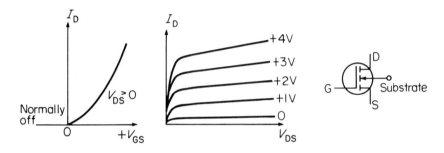

Figure 6.14

n-channel enhancement mode transistor are shown in *Figure 6.14* together with the circuit symbol. On the transfer characteristic the value of the gate-source voltage at which drain current falls to zero is called the threshold voltage V_T. It is that voltage which *just* induces the *n*-channel at the surface of the *p*-type substrate. The curve does not cross the I_D axis. The enhancement FET is a normally OFF device.

> (9) Sketch on appropriately marked axes the characteristic curves of a *p*-channel enhancement type MOSFET.

Figure 6.13(b) shows the construction of an *n*-channel depletion mode MOSFET. Here the *n*-channel is introduced during manufacture by *n*-type doping of the surface layer of the substrate between the *n*-type source and drain regions. The construction is then essentially similar to the JUGFET except that the gate is insulated from the channel by the oxide layer. As a result, current flows from source to drain with zero gate voltage when a positive voltage is applied at the drain. If the gate is made *negative* with respect to the channel, the *n*-channel width is reduced (or depleted) as electrons are expelled from its interface with the oxide dielectric, and drain current decreases. The transistor is then operating in the depletion mode, just as the JUGFET does. Here, however, the gate voltage may be made positive, in which case the channel width is enhanced and drain current increases. Hence

74 *The unipolar transistor*

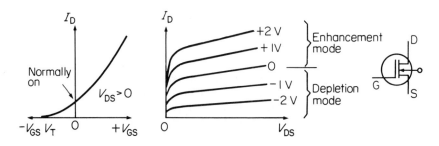

Figure 6.15

this type of MOSFET will operate in either the depletion or enhancement modes. The depletion FET is a normally ON device.

Figure 6.15 shows the output and mutual characteristics of a transistor of this kind. Notice that the threshold voltage now has a negative value (as for the JUGFET) and that the transfer curve crosses the I_D axis, where $V_{GS} = 0$ and $I_D = I_{DSS}$. The symbol for the depletion MOSFET is shown to the right of the diagram. The channel is now represented by a full line.

PROBLEMS FOR SECTION 6

(10) Sketch the construction and describe the principle of operation of a junction field-effect transistor. Explain what is meant by pinch-off and show its effect on the drain characteristics.

Draw a circuit diagram of an automatically biased junction FET used in a Class-A amplifier in the common-source mode. (C. & G.)

(11) Describe the principle and operation of either of the following insulated-gate field effect transistors: (1) depletion type, (ii) enhancement type. Sketch the symbols for these devices and draw typical gate/drain characteristic curves for both types.

(12) Define the parameters: (a) mutual conductance g_{fs}; (b) drain slope resistance r_d; (c) amplification factor μ. Given the relevant characteristic curves of a FET, how would you determine these parameters?

(13) A FET has a transconductance (mutual conductance) of 3 mS. If the gate-source voltage changes by 1 V, what will be the change in the drain current? What assumption have you made in obtaining an answer?

(14) When the drain voltage of a FET is reduced from 20 V to 10 V the drain current falls from 2.5 mA to 2.35 mA. Assuming the gate-source voltage remains unchanged, what is the drain slope resistance of the FET?

(15) The following values are taken from the linear portions of the static characteristics of a certain FET:

V_{DS}	12	12	8	volts
I_D	6.2	2.9	5.9	mA
V_{GS}	−1	−2	−1	volts

Estimate the parameters r_d, g_{fs} and μ.

(16) Complete the following statements:
(a) In general terms, FET's have properties most similar

to .....
(b) A JUGFET can operate only in the ..... mode.
(c) The V_{GS} polarity for an enhancement mode FET is dependent upon .....
(d) I_{DSS} is the drain current which flows when V_{GS} is ....
(e) In a JUGFET majority carriers flow from ..... to .....

(17) A FET has parameters g_{fs} = 2.0 mS, r_d = 50 kΩ. What is its amplification factor? If this FET is used as a common-source amplifier with a drain load of 10 kΩ, calculate the voltage gain.

(18) A FET used in common-source mode has g_{fs} = 2.0 mS, r_d = 100 kΩ. What value of load resistor will provide a voltage gain of 28 dB?

(19) In the circuit of *Figure 6.9*, V_p = −2 V and I_{DSS} = 2 mA. It is desired to bias the circuit to a drain current of 1.1 mA. If the drain supply is 15 V, estimate (a) V_{GS}; (b) g_{fs}; (c) source resistor R_S; (d) the required value of R_L to provide a voltage gain of 10.

(20) The output characteristics of a FET are given in the table. Plot these characteristics and determine the drain-source resistance from the characteristic for V_{GS} = −2 V.

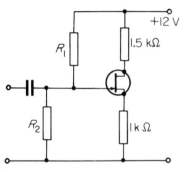

Figure 6.16

	I_D (mA)			
V_{DS} (V)	V_{GS} = 0	−1.0	−2.0	−3.0
2	5.6	4.15	3.0	1.90
6	7.0	5.25	3.62	1.91
9	8.1	6.0	4.1	1.92

Use the curves to estimate the g_{fs} of the FET for V_{DS} = 5 V.

(21) In the common-source amplifier of *Figure 6.16*, the quiescent conditions are such that V_{GS} = −1.0 V and the drain current I_D = 4 mA. Assuming the input resistance of the FET itself is infinite, obtain values of R_1 and R_2 such that the effective input resistance of the amplifier is 0.75 MΩ. What will be the approximate gain of the stage if g_{fs} = 5 mS? (*Hint:* R_1 and R_2 are effectively in parallel with the input terminals).

(22) The equation for the drain current I_D of a FET is

$$I_D = 7.5 \left[1 - \frac{V_{GS}}{-2.5}\right]^2 \text{ mA}.$$

for a V_{DS} of 9 V. Find I_D for each of the following values of V_{GS}: 0, −0.5, −1, −1.5, −2 and −2.5 V. Plot the mutual characteristic for this FET and estimate g_{fs} when V_{GS} = −1.5 V.

(23) A FET is used as a common-source amplifier with a parallel tuned resonant ciruit as the drain load. The tuned circuit consists of an inductor of 400 μH and resistance 20 Ω in parallel

with a 300 pF capacitor. Calculate the resonant frequency.
If the transistor parameters are g_{fs} = 1.5 mS, r_d = 20 kΩ, calculate the voltage gain of the amplifier.

7 Stabilised power supplies

Aims: At the end of this Unit section you should be able to:
Identify the requirements for maintaining a constant voltage output across a load.
Understand the principles of shunt and series stabilisation.
Sketch the block and circuit diagrams of simple series stabilised power supplies using comparator methods.
Explain the operation of such stabilised power supplies.

Direct current power supplies for electronic equipment may, in many cases, be obtained from cells or batteries. Pocket calculators, for example, are nearly always designed to operate from a number of dry cells and in a large number of cases, provision is made for these cells to be periodically recharged, so avoiding the inconvenience of replacement for considerable periods. Flashlamps and portable transistor radio receivers along with car radio and cassette players are other familiar examples where battery supplies are used.

However TV receivers are not usually operated by batteries (apart from those designed for this purpose) and neither would it be very economical or particularly practicable to operate, say, large computer systems or high-power transmitters, or even the domestic hi-fi equipment, from battery supplies. In all these cases, the power is obtained from rectified a.c. mains supplies.

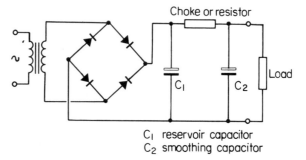

C_1 reservoir capacitor
C_2 smoothing capacitor

Figure 7.1

Rectification of the alternating supply is a relatively simple matter. Full-wave or bridge rectifier circuits used in conjunction with suitably wound transformers will provide us with a direct voltage output, and if this output is smoothed and filtered by a choke-capacitor or resistance-capacitor circuit, a reasonably steady output is available. Such a basic power supply system is shown in *Figure 7.1* and for many pieces of electronic equipment such a supply is perfectly adequate. For more demanding conditions however, for example, in supplying d.c. amplifier or stable oscillators, circuits of this sort are unsuitable. The main reasons for this are:

(i) The output voltage changes if the mains input voltage varies;
(ii) The output voltage changes if the load conditions vary;
(iii) The ripple component in the output which, for 50 Hz supplies, is at a frequency of 100 Hz, increases as the current demands of the load increases.

Such supplies are said to be *unstabilised* against these (usually inevitable) variations. A *stabilised* or regulated power supply, on the other hand, is designed to reduce ripple and ripple variations to a minimum and to provide a stable output voltage regardless of mains voltage or load current variations. In other words, we want a circuit which behaves as closely as possible to the theoretical constant-voltage generator which has negligible internal resistance.

There are two types of regulated power supplies—shunt regulators and series regulators. We will investigate these in turn.

SHUNT REGULATION

The simplest form of shunt regulation is the single zener diode connected as shown in *Figure 7.2*. The regulator element (the zener) is wired in parallel with the load and is fed current by way of series resistor R_S. The elementary theory of zener stabilisation was covered earlier in this book.

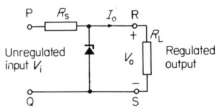

Figure 7.2

The zener diode is connected into the circuit so that it is reverse-biased. At a particular voltage for which the diode has been designed, breakdown occurs at the junction and thereafter the voltage across the diode (V_z) remains substantially constant, irrespective of the current flowing through the diode. This reverse current flow would be destructive in an unprotected diode, but here it is limited to a safe value by resistor R_S so that the power dissipation of the zener is not exceeded. Provided that the voltage across the zener does not fall below the breakdown value, the diode behaves as a current reservoir.

Referring to *Figure 7.2*, it is not difficult to understand how the zener diode provides a constant output voltage V_o at terminals R and S in spite of variations in either the supply voltage V_i at terminals P and Q or in the load current I_o flowing through R_L. Suppose the d.c. input voltage at P-Q increases for some reason. Then the current through the zener diode increases but as the voltage across its terminals remains constant, the increase in voltage appears across R_S. Conversely, if the d.c. input at P-Q decreases, the zener diode surrenders the extra current and the voltage across R_S falls. Hence the input variation is *absorbed by resistor* R_S and the wanted output voltage at R-S remains constant.

Suppose now the load current I_o increases for some reason. The zener current will decrease by the same amount. Similarly, if the load current decreases, the zener current will increase by the same amount.

Stabilised power supplies 79

This time the zener takes up any excess current and sheds any current difference demanded by the load, so acting as a current reservoir while maintaining a constant voltage at the output terminals. General purpose zener diodes are readily available in a range of breakdown voltages running from about 2.5 V up to some 200 V, with power ratings ranging from 400 mW up to 20 W and more.

SERIES REGULATION

In the series form of regulator, the control element is connected in series with the supply lead to the load. If the input voltage V_i or the load current I_o change, the voltage across the control element will adjust itself in such a way that the output voltage will remain constant.

A basic form of series regulator is shown in *Figure 7.3*. This circuit is essentially a current amplifier used in conjunction with a zener diode which acts as a fixed voltage reference V_{ref}. Strictly, we have a shunt regulator (the zener diode) acting through a series element (the current amplifier).

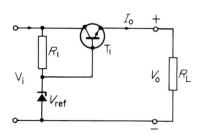

Figure 7.3

The power transistor T_1 is capable of supplying a relatively large current. Its base is held at a constant voltage by the zener diode, supplied with current through resistor R_S exactly as we have already described. This transistor then acts as a current amplifier, so enabling the circuit to provide much higher currents than can be handled by the zener diode alone. If, for example, the current gain of the transistor is 50, then a base current of 20 mA will provide an output current of 1 A. Such a base current is now the effective zener load.

The circuit regulates as follows: suppose the base-emitter voltage is 0.75 V when the transistor is delivering a current of 1 A, then the output voltage will change by 0.75 V when the current drawn from the circuit changes by 1 A. Hence the resistance of the supply as seen at the output terminals is $\delta V_o / \delta I_o = 0.75/1 = 0.75 \Omega$.

Although such a circuit is often adequate for supplying relatively constant load currents at a fixed voltage, some degree of sophistication is called for in more stringent cases.

COMPARATOR METHODS

In *Figure 7.4* an *error amplifier* is used to *compare* the output voltage V_o with a reference voltage V_{ref} and controls the current flowing in the series regulator in such a direction that the error is minimised.

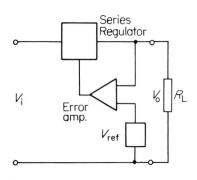

Figure 7.4

A practical form of this circuit is shown in *Figure 7.5*. Transistor T_1 is the series regulator, transistor T_2 with collector load R_2 is the error amplifier. The emitter of T_2 is held at a constant voltage determined by the zener diode fed, in the usual way, through R_1. Resistors R_3 and R_4 feed back a portion of the output voltage to the base of T_2 and are selected so that, in addition, the base of T_2 is positive with respect to its emitter. Hence T_2 is conducting and there is a voltage drop across its collector load R_2. The output at the collector feeds directly to the base of T_1.

Suppose now an output voltage change occurs. This change is sensed at the junction of R_3 and R_4 and hence at the base of T_2. Since the emitter potential is fixed, the base variation will either increase the current through T_2 or decrease it, according as the output voltage variation is an increase or a decrease respectively. The variation, in other words, is compared with the zener reference, and the transistor senses in which direction its base potential has shifted relative

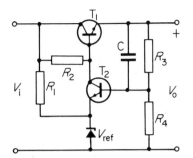

Figure 7.5

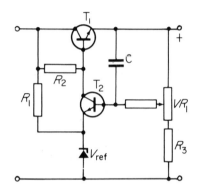

Figure 7.6

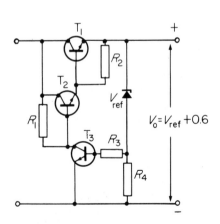

Figure 7.7

to that reference. The change is amplified by T_2 and is fed to the base of T_1 where it adjusts the output current flowing through T_1 in such a way that the original output variation is minimised. You will probably spot that this is an example of negative-feedback, and the circuit can be analysed in terms of negative-feedback theory. This is, however, beyond your immediate course requirements.

In circuits of this sort, the gain of T_2 should be as large as possible, and T_1 must be capable of dealing with the full output current and generally with high power. The voltage drop across T_1 must also be large enough at all times to ensure that it conducts at all times. You may wonder about the capacitor C connected across R_3. This is included to enable the circuit to cope with rapid voltage variations, for R_1 is then effectively short-circuited by C and the full output variation is imposed on the base of T_2. For slow output variations, the output change is attenuated by the potential divider R_3-R_4; since the base and emitter of T_2 are practically at the same potential; the attenuation is in fact

$$\frac{V_o}{V_{ref}} = \frac{R_3 + R_4}{R_4}$$

From this relationship it can be seen that

$$V_o = \left[\frac{R_3 + R_4}{R_4}\right] V_{ref} = \left[\frac{R_3}{R_4} + 1\right] V_{ref}$$

so that V_O can be adjusted by manual variation of the ratio R_3/R_4.

A circuit with adjustable output voltage control is shown in *Figure 7.6*. Potentiometer VR_1 replaces the fixed divider chain and permits a limited range of control, but the output voltage can never be less than that of the reference. If it is required to control the output voltage down to zero, a negative supply rail has to be provided.

(1) Why cannot the output voltage be reduced below the reference level?

(2) What would happen to the output voltage, if anything, if it did drop below the reference level?

Example (3) *Figure 7.7* shows a series stabilised regulator using two transistors, T_2 and T_3, in the amplifier feedback path. Deduce how this circuit operates to maintain a constant output voltage level.

This circuit looks rather different from the previous ones. The zener diode, for example, is on the 'other side' of T_1 and its feed resistor R_4 is returned to the negative rail. However, taking the circuit a step at a time soon reveals that there is quite a lot we will recognise. Clearly, it is immaterial whether the zener has its series resistor on the positive side of the supply or the negative. The zener will break down at its rated voltage V_z and thereafter it will behave as a constant voltage source. Ignoring R_3 for the moment, the output voltage will be the sum of V_z and the base-emitter voltage of T_3 (about 0.6 V). So the zener rating determines the order of output voltage we are going to get.

We have already noted that the resistor in series with a zener takes up any voltage changes occurring across the combination; hence if the output voltage tends to change, the potential across R_4 will change accordingly and the base input to T_3 also changes. The output of T_3 drives the base of T_2; this transistor is in emitter-follower connection and the load impedance output at the emitter controls the series regulator T_1 forming the emitter load of T_2.

Check the polarity variation round the circuit for yourself and verify that the control exercised by T_1 is in the proper direction to reduce the original output voltage change.

Resistor R_2 is included to shunt away the collector-base leakage current of T_1, while R_3 is selected to limit the base current of T_2 in the event of a fault developing.

PROTECTION

There is no inherent short-circuit protection for the series regulator as there was for the shunt circuit. If the output terminals of the series regulator were accidentally short-circuited or if an excessive current was inadvertently drawn by the load, the very large resulting current would, in all probability, cause excessive power dissipation in the control transistor and burn it out.

Protection can be provided by additional circuitry as shown in *Figure 7.8*. When an excessive current flows in the current sensing resistor, the voltage developed across it becomes great enough to switch on the overload detector. This then takes control away from the error amplifier and biases the series regulator back to a safe current condition, usually arranged to be just above the normal maximum current limit.

Again, a practical form of such a system is shown in *Figure 7.9*. This appears even more complicated, but we have the circuit of *Figure 7.7* with one additional transistor, T_4. This forms the overload detector. When the current drawn by the load exceeds a certain level, the voltage across R rises sufficiently from the base-emitter voltage of T_4 to exceed its switch-on value of 0.6 V. T_4 then conducts and diverts base current from T_3, hence limiting the output current to a safe, preset value. The point at which T_4 switches on (in terms of the voltage across R) is decided by the setting of VR_1 so a range of current limiting values is possible. The current sensing resistor R is usually of 1 to 2 Ω resistance, with the current setting control a few hundred ohms.

This is as far as you need to go into the study of basic voltage regulator systems. We have covered only the most elementary methods of such control, but the principles involved are common to the most sophisticated circuits.

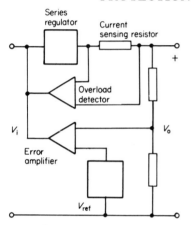

Figure 7.8

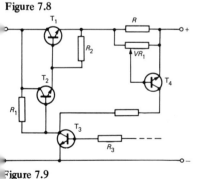

Figure 7.9

PROBLEMS FOR SECTION 7

(4) Differentiate between an unstabilised and a stabilised d.c. power supply.

(5) A 20 V stabilised supply is required from a 40 V unstabilised d.c. source. A 20 V zener diode is to be used for this purpose, having a power rating of 1.5 W. Find the required value of the series resistor.

(6) A zener diode is to provide a 16 V stabilised output from a 20 V unstabilised supply. The load resistor R_L = 2 kΩ and the

82 Stabilised power supplies

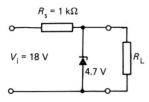

Figure 7.10

zener current I_z = 8 mA. What series resistor is required and what power will be dissipated in it.

(7) In the circuit of *Figure 7.10*, what is the maximum power dissipation in the zener diode? What is the greatest load current that can be drawn from this circuit if the minimum zener current is 3 mA?

(8) The slope resistance of the zener of the previous example is 20 Ω. What will be the change in the no-load output voltage when the input voltage falls to half its present value?

(9) State the requirements for maintaining a constant voltage output across a load. Explain the operation of the power supply unit shown in *Figure 7.11*.

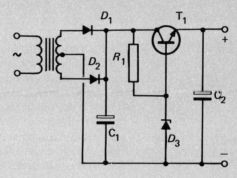

Figure 7.11

8 Logic circuits

Aims: At the end of this Unit section you should be able to:
Understand the elementary rules of circuit logic and the construction of logical functions.
State the logical functions of AND, OR and NOT circuits.
Construct truth tables for AND, OR and NOT functions.
Recognise the Boolean symbols and the circuit symbols for AND, OR and NOT circuits.
Prove the equivalence of two logical expressions by the use of truth tables.

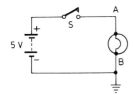

Figure 8.1

In the simple electrical circuit of *Figure 8.1* there are two possible states or conditions in which the circuit may be set: either switch S is closed and the lamp across terminals A and B is ON, or switch S is open and the lamp is OFF. This is an elementary *two-state* system. With such a system it is easily possible to transmit information from one place to another by using any prearranged code where the letters of the alphabet are represented by ordered arrangements of the ON and OFF states. Morse code itself is a two-state language, but here ON and OFF are replaced by SHORT and LONG; the signal pulses representing the letters are either dots or dashes or arrangements of these two possible duration states.

From a consideration of such elementary two-state ideas as these, it may seem a big step to computers and electronic control systems, but these also operate in two-state coding languages, and what appear to be extremely complex electronic devices are often actually complex only in the repetitive use of such basic circuits as that of *Figure 8.1* and one or two others we are going to meet on the following pages. In this section we shall be interested in what are called *logical circuits* and their associated so-called *logical equations*, circuits which operate in one or other of two possible conditions. Whether the conditions are simply referred to as ON and OFF, as in the lamp circuit, or as HIGH and LOW, UP and DOWN, PLUS and MINUS, TRUE and FALSE, to list only a few of the possibilities, is immaterial, but the number symbols 0 and 1 are probably best suited for our particular purpose and these are the symbols we shall normally use.

Logic itself, in its normally accepted sense, treats of the validity of thought and reason, truth and falsity being its two basic propositions. It may seem a far cry from a philosophy of this kind to the design of electronic circuits, but the complete symbolisation (putting into an algebraic form) of logic with its fundamental two-state foundations of truth and falsity has been made over the past hundred years, principally by the pioneering work of the mathematician George Boole (1815–1864), and applied to electronic systems in particular over the past 30 years or so.

Why are we particularly interested in two-state systems? Simply because so many electrical and magnetic devices have essentially two stable states, ON and OFF.

POSITIVE AND NEGATIVE LOGIC

There is one point to take care of before we go any further. Remember, we have decided on the use of the symbols 0 and 1 to represent each of two possible circuit states. However, if we simply leave things at that, an ambiguity is soon going to catch up with us. Go back to *Figure 8.1* and consider the voltage across the lamp terminals A and B, B being connected to earth, say. When the switch is open there will be 0 V at A with respect to earth and when the switch is closed there will be +5 V at A with respect to earth. The two possible states of the output therefore are 0 V and +5 V, each with respect to earth or terminal B. To which of these conditions shall we assign either our 0 or 1? It seems commonsense to assign 0 to the 0 V output state and 1 to the +5 V output state, but there is no reason why we should not choose the alternatives.

Suppose now the battery is reversed. This time the two outputs are going to be 0 V and -5 V with respect to earth. Again, we can choose 0 to represent the 0 V state and 1 to represent the -5 V state, but there is no reason why we should not choose the alternatives.

This sounds all very confusing, and as there are no definite rules about the problem it might seem that there are going to be difficulties ahead. However, we will apply the following conventions:

1. *Positive logic* labels the more positive voltage level as the logic state 1 and the other voltage level as the logic state 0.
2. *Negative logic* labels the more negative voltage level as the logic state 1 and the other voltage level as the logic state 0.

For most of our work in this Unit section we will operate in positive logic convention as conforming more to our 'commonsense' view of the two possible states of a system, that is, high level = 1, low level = 0.

RULES OF CIRCUIT LOGIC

Simple switches provide us with a convenient starting point in the study of circuit logic because circuit devices such as relays, diodes and transistors can be switched on or off by simple circuit codes.

Figure 8.2 shows two switches connected in series with a battery (the signal) and a lamp (the output). We shall allocate the symbol 0 to represent an open (unoperated) switch and a 1 to represent a closed (operated) switch. Further, we shall represent an output voltage (lamp on) by 1 and no output voltage (lamp off) by 0. All these allocations conform to positive logic. Now suppose we wish to know the switch positions required if there is to be an output from the lamp. No problem here — obviously the lamp will light only if both switch A and switch B are closed. If either A or B or both are open (0), the output voltage will be 0. We accordingly call such a series circuit a *logical AND circuit*: there is an output (1) only if A = 1 and B = 1. Call the output F, then

$$A \text{ and } B = F$$

The symbol for AND is a dot (as used in the algebraic symbolism for multiplication), so we may write

$$A \cdot B = F$$

We can gather the information deduced from this circuit in the form of a table. This is called a *truth table*, and this is what it looks like:

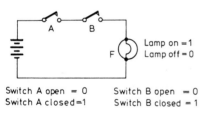

Switch A open = 0 Switch B open = 0
Switch A closed = 1 Switch B closed = 1

Lamp on = 1
Lamp off = 0

Figure 8.2

Logic circuits 85

Table 8.1

A	B	F	
0	0	0	⟶ A and B are off, no output
0	1	0	⟶ A is off and B is on, no output
1	0	0	⟶ A is on and B is off, no output
1	1	1	⟶ A is on and B is on, an output

Notice that there are four rows to the table. There are two switches each with two possible positions, so there are $2^2 = 4$ ways of arranging the switching. The notes on the right of the table show how the A and B rows are built up.

Check this table carefully to make certain you understand the way it has been set out: the A and B switch states have been put down in all four possible ways, and only if A and B are present together is there an output F.

We note: series connected switches represent logical AND functions.

Let the switches now be connected in parallel as shown in *Figure 8.3*. This time the lamp will light (1) if either A or B is closed (either = 1) or both A and B are closed (both = 1). If both A and B are open (0) the output F will be 0. As before there are four possible switch arrangements, but if either A = 1 or B = 1, or if A = B = 1, the circuit will be closed and the output will be 1. Such a parallel switch circuit represents a *logical OR circuit*.

The truth table for the logical OR function is as follows:

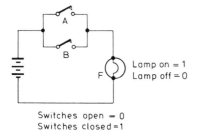

Figure 8.3

Table 8.2

A	B	F
0	0	0
0	1	1
1	0	1
1	1	1

The A and B states have been set down in all four possible ways; only if A or B, or both A and B are present is there an output F, symbolised in the lower three rows of the table.

From the table, the logical equation can be deduced as

$$A \text{ or } B = F$$

The symbol for OR is +, although the symbol v, now superseded, will be found in many books dealing with logic. Hence

$$A + B = F$$

(or A v B = F in the superseded symbolism).

We note: parallel connected switches represent logical OR functions.

Follow the next two worked examples carefully to familiarise yourself with these new and probably strange concepts.

Example (1). How many rows would there be in a truth table for a circuit containing *n* switches?

Since the variable quantities of logic have two states, or values, a certain number of binary variables taken together will produce a finite number of possible combinations. For example, if our variables are two switches each of which may be either ON or OFF, both switches taken together give four possible combinations: these we have seen in the earlier pages as OFF-OFF (0,0), OFF-ON (0,1). ON-OFF (1,0) and ON-ON (1,1). If we used three switches, each of which could be either ON or OFF, we would get eight possible combinations, for each of the above cases for two of the switches could be combined with the two positions for the third switch. Four switches would result in sixteen combinations and so on. The relationship is not difficult to spot as they go up in powers of 2: the number of possible combinations of the variables (switch positions) is 2^n where *n* is the number of switches.

Example (2). Draw circuit diagrams and draw up truth tables which represent the following logical equations:

(i) A.B.C = F;   (ii) A.(B + C) = F.

(i) This reads as 'A and B and C = F'. So there is an output F when A, B and C are simultaneously present. The circuit is clearly a series arrangement of three switches (*Figure 8.4*).

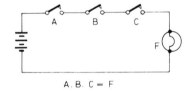

A.B.C = F

Figure 8.4

With three switches, each with two possible states, there are 2^3 = 8 arrangements of the switch positions. Writing these in order under A, B and C headings gives the first three columns of the truth table (remember switch OFF = 0, switch ON = 1), the only condition for an output (1) at F is when switches A, B and C are all closed together. This corresponds to the last row in the table, so here F = 1. All other conditions for A, B and C give F = 0.

Table 8.3

A	B	C	F
0	0	0	0
0	0	1	0
0	1	0	0
0	1	1	0
1	0	0	0
1	0	1	0
1	1	0	0
1	1	1	1

(ii) This equation reads as 'A and (B or C) = F'. The circuit has to contain three switches A, B and C, and an output F will be obtained if A together with B and C are operated. The OR function represents a parallel arrangement of switches B and C, while the AND function puts switch A in series with them. The circuit is then assembled as shown in *Figure 8.5*. Satisfy yourself that this circuit will indeed perform the requirements stated in the given logical equation.

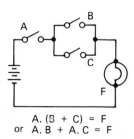

A.(B + C) = F
or A.B + A.C = F

Figure 8.5

For the truth table, there are once again three switches having eight possible contact arrangements, so the table has eight rows identical with those given in the previous example. An output F is only obtained, however, if at least A = 1, and either B or C or both B and C together are 1. The table follows:

Table 8.4

A	B	C	F
0	0	0	0
0	0	1	0
0	1	0	0
0	1	1	0
1	0	0	0
1	0	1	1
1	1	0	1
1	1	1	1

As A is 0 in the first four rows, F must also be 0, irrespective of what B and C happen to be. The fifth row has A = 1, but both B and C are 0, hence F is 0. Only the last three rows satisfy the condition of A present with B or C, or B and C together present at the same time.

Suppose we had 'multiplied out' the bracketed term in the equation on this example, just as though it had been an ordinary algebraic expression. We should have obtained.

$$A(B + C) = A.B + A.C = F$$

If you compare this alternative against the truth table, you will find that the conditions are satisfied, so the bit of algebraic manipulation seems justified: 'A and (B or C)' is exactly the same as A and B or A and C'. This is quite general: with a few exceptions which are peculiar to logic functions, all the ordinary rules of algebra can be applied.

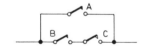

Figure 8.6

Try the next problems yourself.

(3) Draw a circuit diagram and set up the truth table applicable to the logical function A + B + C = F.

(4) Represent the circuit shown in *Figure 8.6* in the form of a logical equation and a truth table.

(5) How many switching combinations are possible with the circuit of *Figure 8.7*? Write down a logical equation representing this circuit and draw up a truth table.

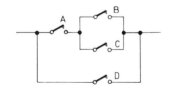

Figure 8.7

NEGATION

In logic, if one position of a two-position switch is A, the other position is not-A, symbolised as an A with a bar over it, $\overline{A}$. In a circuit we can represent negation by a switch that is normally closed as shown in *Figure 8.8*. In this case the unoperated switch represents the 0 condition, while operating the switch sets it to the 1 condition. Output voltage is present (F = 1) when the switch is normally closed (A = 0). The truth table for this circuit follows:

Figure 8.8

Table 8.5

A	F
1	0
0	1

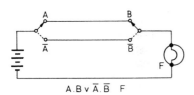

A.B ∨ Ā.B̄   F

Figure 8.9

Example (6). A lamp is controlled by two change-over switches as shown in *Figure 8.9*. Set up the logical equation and truth table for this circuit and discuss the circuit operation in logic equation form.

Each switch has a position we shall call A and B respectively (shown in the full lines). When the switches are moved to their second positions, these become $\bar{A}$ and $\bar{B}$ respectively (broken lines). Again, there are four possible combinations of the switch positions.

Since there is an output (lamp on) for the positions A and B, and also for the positions $\bar{A}$ and $\bar{B}$, we can write

$$A.B = F \text{ or } \bar{A}.\bar{B} = F$$

There is no possibility of a parallel connection, both switches must be in series for the light to operate. In other words, since the switches each have only two positions and both must be the same, there are only two possible paths for the lamp and battery to be connected: A.B is one and $\bar{A}.\bar{B}$ is the other. Hence, by combination, the lamp is on (F = 1) if A and B or not-A and not-B are actuated. Hence

$$A.B + \bar{A}.\bar{B} = F$$

which is the required logical equation.

The truth table is now easily deduced as follows:

Table 8.6

A	B	F	
1	1	1	⟶ A.B = F
1	0	0	
0	1	0	
0	0	1	⟶ $\bar{A}.\bar{B}$ = F

Example (7). Draw circuits representing the logical equations

(i) $A.\bar{B} = F$; (ii) $A.\bar{B} + \bar{A}.B = F$.

(i) This equation asserts that there is an output F (1) whenever A and not-B ($\bar{B}$) are present together. Electrically, the equation can be represented simply by a series circuit containing a normally open switch A and a normally closed switch B, as shown in *Figure 8.10*. The circuit is completed (lamp on, F = 1) whenever switch A is placed in its closed position (1) and switch B is left in its normally closed position (0).

(ii) This equation asserts that there is an output F whenever A and not-B or not-A and B are present. Electrically, the equation can be represented by two sets of series connected switches (A.$\bar{B}$ and $\bar{A}$.B), both sets being connected in parallel. The first set has

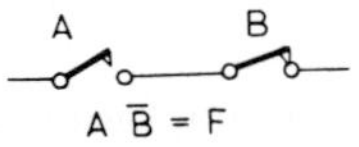

A $\bar{B}$ = F

Figure 8.10

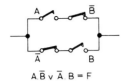

Figure 8.11

B normally closed (for $\bar{B}$) while the second set has A normally closed (for $\bar{A}$). The circuit is shown in *Figure 8.11*.

TESTING FOR LOGICAL EQUIVALENCE

Figure 8.12(a)

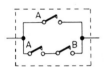

Figure 8.12(b)

Suppose we have two boxes. One contains the switching circuit shown in *Figure 8.12(a)* and the other the circuit shown at *(b)*. The logical equation for circuit *(a)* should be no problem for you; it is

$$A = F$$

For circuit *(b)* you should be able to deduce that

$$A + (A.B) = F$$

As usual, the F simply signifies that we get an output signal when the switches are operated in the manner indicated by the equations. If the two switches marked A seem confusing, remember they are ganged together and so operate together. Now since either equation represents the same circuit condition, we should be able to say that

$$A + (A.B) = A$$

Clearly, this statement is true: if switch A is operated, the circuit is completed. The presence of switch B is irrelevant. So, too, is switch A in series with B. Only the top switch A is strictly necessary, hence the circuit at *(b)* is exactly equivalent to the simple circuit at *(a)*, and so the latter can replace it, thereby saving two switches in the process.

The equivalence of two logical equations is a matter of great importance in the design of computers, since the ability to reduce a complicated circuit to a simpler one doing exactly the same job, enables the designer to use the minimum number of circuit elements (say integrated circuits) to achieve a certain result. Can we prove the logical equivalence of two equations without recourse to circuit diagrams? The answer is yes, and there are several ways of doing it. Only one method will concern us at this stage and this is the method of proof by the use of truth tables.

A truth table is constructed for each of the expressions we are comparing for equivalence (or otherwise). If the two expressions have the same truth value (0 or 1) for each case in the truth table, then the expressions are equivalent. If at any point the truth values differ, then the expressions are not equivalent. The method is best illustrated by worked examples, so follow the next examples carefully.

Example (9). Prove, using truth tables, that $A + (A.B) = A$.

This is the case we previously looked at in terms of switching circuits. To prove the statement, we draw up truth tables as follows:

A	B	A.B	A + (A.B)
0	0	0	0
0	1	0	0
1	0	0	1
1	1	1	1

The first two columns, as usual, represent all possible arrangements of the quantities A and B. The third column is the logical AND function A.B. The fourth column is constructed in accordance with the logical OR function. We compare the A column values with the A.B column values: a 1 is placed in the fourth column when either the A column *or* the A.B column (or both) have the value 1, and a 0 is placed in the column when the value of both the A and the A.B columns is 0.

Comparing the fourth column, i.e. A + (A.B) with the first column, A, they are seen to be identical, hence the statement A + (A.B) = A is shown to be true.

Example (10). Prove that $A + \bar{A}.B = A + B$, and illustrate the equivalence in terms of switching circuits.

As before, we draw up a truth table for both expressions and compare them.

A	B	$\bar{A}$	$\bar{A}.B$	$A + \bar{A}.B$	$A + B$
0	0	1	0	0	0
0	1	1	1	1	1
1	0	0	0	1	1
1	1	0	0	1	1

The third column ($\bar{A}$) is the negation of the first column (A). The fourth column is the logical AND function for $\bar{A}.B$ and has the value 1 only when both $\bar{A}$ and B are simultaneously 1, i.e. the second row condition. The fifth column is the logical OR function and has the value 1 when either the A column or the $\bar{A}.B$ column (or both) have the value 1. The last column is the OR function A + B.

Comparison of the fifth column with the sixth shows them to be identical, hence the statement $A + \bar{A}B = A + B$ is true.

Figure 8.13(a) shows the circuit representation of $A + \bar{A}B$, and *(b)* shows the representation of A + B. These should be identical in circuit function. In *(a)* the circuit is completed if A is closed (note that $\bar{A}$ then opens) or if A is left open (so that $\bar{A}$ remains closed) and B is closed, i.e. $A + \bar{A}.B$ is satisfied. So all that is *necessary* for circuit completion is the closure of either A or B; switch $\bar{A}$ is redundant to the operation. Circuit *(b)* representing A + B is then identical to circuit *(a)*.

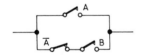

Figure 8.13(a)

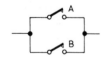

Figure 8.13(b)

LOGICAL RELATIONS AND ALGEBRA

Although the plus (+) and multiplication (.) signs used in ordinary algebra are not to be confused in any way with the OR (+) and the AND (.) signs associated with logical expressions, most rules for the manipulation of logical algebra follow those of ordinary algebra. There are some surprises which only experience will overcome, but a convenient way of looking at logical algebra is to visualise the symbols and their interconnections in terms of simple basic electrical circuits, in the way we have been doing in the earlier parts of this chapter. First of all, fix the following equivalents in your mind:

0 is equivalent to an *open* circuit
1 is equivalent to a *closed* circuit

Logic circuits 91

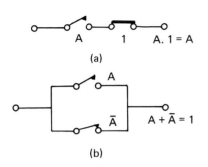

Figure 8.14

A (or B or C etc.) is equivalent to a switch which may be closed (=1) or open (=0)
$\bar{A}$ is equivalent to a switch which may be closed (=0) or open (=1).

For example, A.1 is equivalent to a switch A in series with a closed circuit, see *Figure 8.14(a)*. Since the closed circuit is always closed, the function of the circuit depends only upon the state of A. Hence A.1 = A.

Again, A + $\bar{A}$ is equivalent to two parallel switches, one closed (A) when the other is open ($\bar{A}$). Whatever the setting of the switches, the circuit is always closed via one of them. Hence A + $\bar{A}$ = 1. This is rather like saying that the result of a toss of a coin will result in either HEADS or NOT-HEADS: you are bound to be correct! See *Figure 8.14(b)*.

Table 8.7 summarises these basic relationships in logical algebra in both meaning and circuit interpretation.

Two or More Variables In *Table 8.7* we may take the symbols 1 and 0 as 'constants' in the sense that the closed and open circuits which they respectively represent are *permanent*. The symbol A (or B or C, etc.), however, is a *variable* in that it may adopt either of the two possible states 0 or 1.

Logic	Meaning	Circuit
0.0 = 0	Open in series with open is open	
0.1 = 0	Open in series with closed is open	
1.1 = 1	Closed in series with closed is closed	
A.0 = 0	A in series with open is open	
A.1 = A	A in series with closed is A	
A.A = A	A in series with A is A	
A.$\bar{A}$ = 0	A in series with not—A is open	
0 + 0 = 0	Open in parallel with open is open	
0 + 1 = 1	Open in parallel with closed is closed	
1 + 1 = 1	Closed in parallel with closed is closed	
A + 0 = A	A in parallel with open is A	
A + 1 = 1	A in parallel with closed is closed	
A + A = A	A in parallel with A is A	
A + $\bar{A}$ = 1	A in parallel with not—A is closed	
$\bar{\bar{A}}$ = A	Not(Not—A) = A (double negation)	

Table 8.7

We come now to the rules or theorems relating to logical expressions involving two or more variables. These rules go under the same names as are used in ordinary algebra, and in some cases conform to the rules followed in ordinary algebra. The most important of these are set out in *Table 8.8*.

Table 8.8

1. Commutative rules	$A + B = B + A$ $A.B = B.A$
2. Associative rules	$A + (B + C) = (A + B) + C$ $A.(B.C) = (A.B).C$
3. Distributive rules	$A.(B + C) = A.B + A.C$ $A + B.C = (A + B).(A + C)$
4. Reduction rules	$A + A.B = A$ $A + \bar{A}.B = A + B$
5. De Morgan's rules	$\overline{A + B} = \bar{A}.\bar{B}$ $\overline{A.B} = \bar{A} + \bar{B}$

The first two of these rules need no explanation. The commutative rules tell us that the *order* of the variables in performing OR and AND functions is unimportant; the associative rules tell us that the order of the OR and AND *operations* is unimportant.

The first of the distributive rules is straightforward and follows ordinary algebra in that the bracketed terms can be 'multiplied' out. The second rule does *not* follow ordinary aldebra but indicates that variables or combinations of variables may be distributed in multiplication as well as in addition. Make sure that you grasp this variation.

The two reduction rules (also known as absorption rules) do not follow ordinary algebra either but are easily verified; in the first we factorise the A out of both terms to give

$$A + A.B = A(1 + B)$$
But $\qquad 1 + B = 1$ (see *Table 8.7*)
$$\therefore A + A.B = A.1 = A$$

For the second rule, we expand the left-hand side by using the distributive rule:

$$A + \bar{A}.B = (A + \bar{A}).(A + B)$$
But $A + \bar{A} = 1, \therefore A + \bar{A}.B = A + B$

The final set of rules are known as De Morgan's rules after their originator. They are unique to logical algebra and can be used to negate (or find the inverse) of any logical expression. The rules are best remembered in the form: to negate a logical expression, invert all the variables (exchange A by $\bar{A}$, $\bar{A}$ by A etc.), and replace all OR signs by ANDs and all AND signs by ORs.

It takes practice, as does anything else, to manipulate logical expressions quickly and accurately, using the above rules and theorems, but you should be able to reduce the more simple equations without much difficulty.

Here are a few worked examples as a guide.

> *Example (8).* Verify the following logical equations:
> (a) $AB\overline{C} + ABC = AB$
> (b) $AB + A\overline{B} + \overline{A}B = A + B$
> (c) $\overline{A + B + C} = \overline{A}\,\overline{B}\,\overline{C}$
> (d) $\overline{A}BC + AB\overline{C} + ABC = AB + BC$
> (e) $\overline{\overline{AB} + \overline{A} + B} = A\overline{B}$
>
> (a) Factoring, we get $AB\overline{C} + ABC = AB(\overline{C} + C)$
> But $\overline{C} + C = 1$
> $\therefore AB(\overline{C} + C) = AB$ $\therefore AB\overline{C} + ABC = AB$
> (b) Put $AB + A\overline{B} + \overline{A}B = A(B + \overline{B}) + \overline{A}B$
> $= A.1 + \overline{A}B = A + \overline{A}B$
> But $A + \overline{A}B = (A + \overline{A})(A + B)$ by the distributive rule
> $= A + B$, since $A + \overline{A} = 1$
> $\therefore AB + A\overline{B} + \overline{A}B = A + B$
> (c) This is simply a straight application of De Morgan's rule: on the left-hand side, the *individual* elements A, B and C are not negated, the whole term is; hence we can negate them separately and change all the OR signs to AND signs. This then gives us the term on the right.
> (d) Factoring out the AB out of the last two terms, we get
>
> $\overline{A}BC + AB(\overline{C} + C)$
>
> Since $(\overline{C} + C) = 1$, we get $\overline{A}BC + AB$
> Factoring out the B gives us $B(A + \overline{A}C)$
> $= B(A + C)$ by the distributive rule
> $= AB + BC$
> (e) Apply De Morgan's rule and $\overline{\overline{AB} + \overline{A} + B}$ becomes
> $\overline{(\overline{A} + B)} + \overline{A\overline{B}}$
> $= A + \overline{B}(1 + A)$ by distribution
> $= A + \overline{B}$ since $1 + A = 1$
> $= \overline{A\overline{B}}$ by De Morgan's rule.
>
> Try the following on your own.
> (9) Verify the logical equations:
> (a) $(A + B)(A + B) = A$
> (b) $\overline{A}\overline{C} + B\overline{C} + \overline{A}BC + ABC = B + \overline{A}\overline{C}$ (Hint: factor out twice.)
> (c) $(A + B)(\overline{A} + C) = AC + \overline{A}B$
> (d) $\overline{\overline{AB} + \overline{C + D}} = AB\overline{C}\overline{D}$ (Hint: use De Morgan twice and remember $\overline{\overline{A}} = A$.)

In the next Unit section we shall be looking at yet another way of simplifying logical algebraic expressions.

PROBLEMS FOR SECTION 8

Group 1

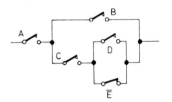

Figure 8.15

(11) What logical functions are represented by the following truth tables:

A	B	F	A	B	F	A	B	F
0	0	0	0	0	0	0	0	1
0	1	1	0	1	0	0	1	0
1	0	1	1	0	0	1	0	0
1	1	1	1	1	1	1	1	1

(12) Draw circuit diagrams and truth tables to represent the following logical expressions:
  (a) A(B + C)
  (b) (A + B).(A + C)
  (c) A(Ā + B)

(13) A circuit which is OR with positive logic will be ...... with negative logic.

(14) Write down the logical expression for the circuit shown in *Figure 8.15*.

(15) Use truth tables to verify the following statements:
  (a) A + B.C = (A + B)(A + C)
  (b) A(B + C) = A.B + A.C
  (c) (A + B)(A.B) = A.B

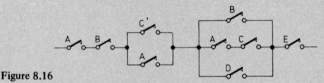

Figure 8.16

Group 2

(16) Write down the logical expression representing the circuit shown in *Figure 8.16*. Verify that your solution can be represented by the simple form F = A.B.E.

(17) What logical expression does the following truth table represent?

A	B	C	F
0	0	0	0
0	0	1	1
0	1	0	1
0	1	1	0
1	0	0	0
1	0	1	1
1	1	0	1
1	1	1	0

Draw the circuit interpretation of the expression.

(18) The expressions 0.0 = 0 and 1.1 = 1 are illustrated respectively at *(a)* and *(b)* in *Figure 8.17*. Using these as a guide, draw the circuit interpretations of
  (a) 0 + 0 = 0   (b) 1 + 1 = 1   (c) 1 + 0 = 1
  (d) A + 1 = 1   (e) A + 0 = A   (f) A.0 = 0
  (g) A.1 = A

(19) Verify the following statements:
  (a) A(Ā + B) = A.B

Figure 8.17

(b) $(A + \bar{B})(A + C) = A + \bar{B}.C$
(c) $\overline{A + \bar{B}} = \overline{A.B}$
(d) $\overline{A + B} = \overline{A}.\overline{B}$

These last two expressions are De Morgan's Rules and you will meet them again in more advanced work. You should make a note that $\overline{A.B}$ means 'not (A and B)'. It is not the same as $\overline{A}.\overline{B}$ which means 'not A *and* not B'.

9 Combinational logic gates

Aims: At the end of this Unit section you should be able to:
Recognise BS and ANSI symbols for the basic logic gates.
Understand the action of diode, RTL and TTL logic gates.
Draw the circuit representation of simple logical expressions.
Derive truth tables for practical gate circuits in combination.
Select from manufacturer's lists the standard commercial TTL and CMOS logic gate families.

ELECTRONIC GATES In the previous section we dealt with logical relationships and their interpretation in the form of simple switching circuits. So that we can use these (and other) relationships in practical circuits it is necessary to have a series of electronic gate elements which will enable the desired logic to be achieved not by hand-operated switches but by purely electronic means. Five basic gates are of interest at this stage: those covering the three logical functions dealt with in the previous section plus two others which can be looked on as combinations of these, namely, the NOT-AND or NAND gate and the NOT-OR or NOR gate. The symbols for these gates, plus the X-OR or EXCLUSIVE OR gate, of which more later, are given in *Table 9.1*. You should become familiar with both the BS (British Standard) symbols and the ANSI (American)

BS	ANSI		
A,B,C → [&] → F	A,B,C → AND shape → F	AND	F = A.B.C
A,B,C → [1] → F	A,B,C → OR shape → F	OR	F = A + B + C
A → [1]o → $\bar{A}$	A → ▷o → $\bar{A}$	NOT	
A,B,C → [=1] → F	A,B,C → XOR shape → F	EX. OR	F = A $\oplus$ B $\oplus$ C

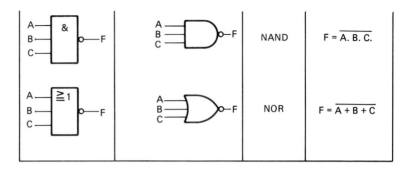

Table 9.1

symbols; in this book we will use a mixture of the symbols to give you familiarity, though the emphasis will be on ANSI.

In the diagrams of *Table 9.1* we have assumed that each gate has three inputs, but any number may be involved in practice. The NOT gate, of course, is simply an inverter and has an output which is the negation of the single input.

Each of the inputs to these gates consists of electrical signals of a binary code: signals which will represent one or other of two possible states, generally voltage levels. We have seen how the 'signals' in the previous section have been represented by ON and OFF switch positions which is the same thing as the corresponding presence or absence of voltage respectively. In many practical applications two voltage levels are used, zero voltage corresponding to 0 in the logical algebra and (typically) +5 volts corresponding to logical 1. There are tolerances and limits to all such voltage levels used in this way, and these will be discussed in due course. For now, we shall assume positive logic with voltage levels of 0 and +5 V.

ELECTRONIC SWITCHING

How can we use diodes and transistors, alone or in combination, as our switching elements in the logical systems already discussed? *Figure 9.1(a)* shows an elementary diode switch. If the input terminals A and B of this circuit are short-circuited (so that the input is zero or logical 0), the diode will be forward biased and a current will flow down through resistor R. The output voltage will then be equal to that across the diode which, for a silicon device, is about 0.6 V. This level comes within the accepted tolerance for low or logic 0. The output is therefore low when the input is low. Suppose now that the input is raised to 5 V or more; the diode will switch off and the output will rise to about +5 V. The output is therefore high if the input is high. The ON-OFF conditions of the diode switch correspond to voltage levels close to zero and +5 V. Notice that the input and output levels correspond, that is, there is no inversion.

We have already mentioned the transistor as a switch in an earlier section but we include it again at this point to make the record complete: *Figure 9.1(b)*. With zero input, the base-emitter junction is reverse biased by $-V_B$, the collector current is zero and the output is about 5 V. A positive input greater than V_B switches the transistor on and a large collector current flows through R_3. The collector potential, and hence the output, falls to a low level, typically 0.2 V. Again the output levels are near zero and 5 V, but this time the input has been inverted. This, of course, is the NOT gate.

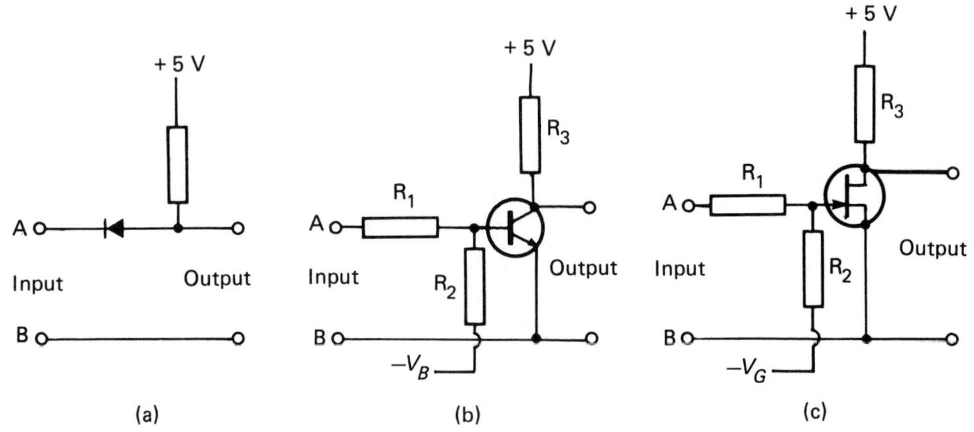

98 Combinational logic gates

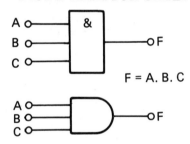

Figure 9.1

The junction FET will behave similarly to zero and 5 V levels applied to the input terminals, see *Figure 9.1(c)*.

We can now apply these basic switching circuits to the design of practical gates.

DIODE-RESISTOR GATES

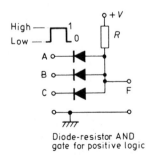

Figure 9.2

Figure 9.3

The logical AND gate receives a number of input signals but does not deliver an output unless all those inputs are at logic level 1. If therefore we designate the input signals as A, B and C, there is an output only if A and B and C are simultaneously present at a high level, that is, F = A.B.C, as *Figure 9.2* shows. For this reason, the AND gate is often referred to as a *coincidence gate*.

We now look at the case of a diode-resistor AND gate, seen in *Figure 9.3*. The magnitude of the positive anode supply voltage V is set to be a few volts higher than the logical 1 level of the input signals, taken to be +5 V. The operation of this gate is relatively easy to follow through.

Suppose all input signals are simultaneously high; then *all* the diodes conduct because of the more positive anode voltage V, and the output is connected to the high level inputs directly through the diodes. Hence the output F is high. The switch analogy in this case is shown in *Figure 9.4(a)*. Assume now that any one of the inputs, A, for example, goes low (zero volts). The cathode of diode A is now low but the anode is high, hence the diode conducts and connects the output to the low level A input. This has the effect of bringing the anodes of diodes B and C also to a low level since all anodes are connected together. Hence the presence of high level conditions at the *cathodes* of B and C in conjunction with the low level anodes causes these two diodes to switch off, and no signals from inputs B and C can reach the output. The output is correspondingly low, as the switch analogy of *Figure 9.4(b)* shows. By exactly the same reasoning (which you should go through in detail for yourself), the output will be low whenever any two of the inputs are low and the third is high or all three are low. Only the simultaneous presence of high level inputs at all three input terminals produces a high at the output, hence this circuit performs as a logical AND for positive logic inputs.

Combinational logic gates 99

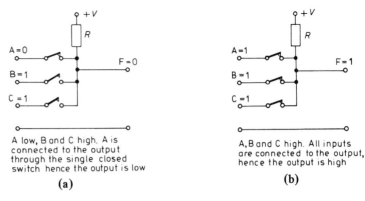

Figure 9.4

Input pulses are not always of the same duration, so it is possible for the AND gate to act as a detector of coincidence among several inputs. A possible train of positive input pulses at points A, B and C is shown in *Figure 9.5*. Only when inputs A, B and C are simultaneously high will there be an output F. From the diagram, this state of affairs occurs only during the time interval t_1 to t_2.

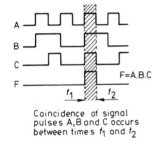

Coincidence of signal pulses A, B and C occurs between times t_1 and t_2

Figure 9.5

(1) Complete the following:
(a) The AND gate is sometimes referred to as a .... gate.
(b) The AND gate behaves like .... connected switches.
(c) The output of an AND gate with one input held low will be at logic .....

Diode-resistor OR Gate

An OR gate, symbolised in *Figure 9.6*, provides an output if any one of its inputs is at logic 1. If therefore we designate the input signals as A, B and C, there is an output high if A or B or C is present, that is, F = A + B + C. At this point we should distinguish between the two forms of OR gate, the inclusive and the exclusive OR. Consider for simplicity a two-input OR gate; if an output is obtained when A = 1 or B = 1 or if *both* are 1 together, the gate is known as an *inclusive* OR. If, however, an output is obtained only when A = 1 or B = 1, but not if both equal 1, the gate is known as an *exclusive* OR. These definitions obtain when any number of inputs are involved. Our earlier analogy of an OR gate to switches connected in parallel was the inclusive case, and the OR circuit truth table (Table 8.2), to which you should refer, illustrates the inclusive case. We shall return to the exclusive OR gate a little later on.

Figure 9.6

(2) Draw the truth table for a two-input exclusive OR gate.
(3) Using a couple of switches, a battery and a bulb, design a simple *exclusive* OR circuit. (Hint: you will have to use change-over type switches, not simple on-off ones.)

Look now at the diode-resistor circuit of *Figure 9.7*. The diode cathodes are returned to a negative voltage point (this is sometimes simply zero level) and the input pulses conform to positive logic. This

100 Combinational logic gates

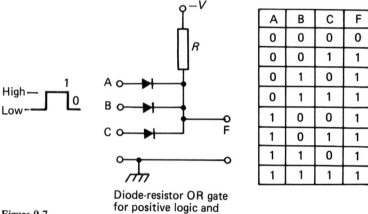

Figure 9.7

Diode-resistor OR gate for positive logic and truth table

circuit will perform the inclusive OR function in which any single high input (or two or all three) will produce a high output. When all the inputs are low, all diodes conduct, the cathodes being more negative than the anodes. The output is then connected directly to the input and so the output is low. This condition relates to the first row of the truth table in the figure. Suppose now that any input, say C, goes high. Diode C will conduct and the output will be high; diodes A and B then have high cathodes but low anodes and consequently switch off. This condition clearly applies for any other input, or combinations of the inputs, being high, hence the truth table can be completed as shown.

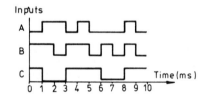

Figure 9.8

(4) How would the truth table shown in *Figure 9.7* be modified if it applied to an exclusive OR gate?

(5) *Figure 9.8* shows a train of input pulses which are applied in turn to three-input (a) positive AND gate, (b) inclusive OR gate, (c) exclusive OR gate. The time scale indicates milliseconds of duration. Between what times are there high level outputs from (i) the AND gate, (ii) the OR, (c) the exclusive OR gate?

(6) A diode–resistor OR gate provides a high output when one or more of its diodes are .... biased.

GATE COMBINATIONS Up to this stage we have explained logic gates in the form of single elements. Using the three basic elements of AND, OR and NOT functions, it is possible to derive arrangement and combinations which represent a great number of logical expressions.

Consider first a two-input AND gate. *Figure 9.9* shows four possible output expressions for the logical AND by the inclusion of the NOT gate in three of the circuits. The outputs are respectively A.B, $\overline{A}$.B, A.$\overline{B}$ and $\overline{A.B}$. Make sure that you understand how each of these outputs comes about.

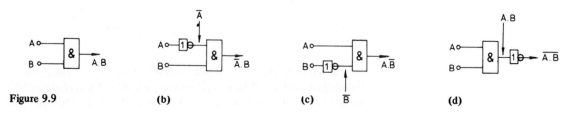

Figure 9.9   (b)   (c)   (d)

This is best done by considering the 'intermediate' input states shown arrowed; for example, in case (b) input A is inverted to $\bar{A}$ and this together with B forms the input to the AND gate. The final output is then $\bar{A}.B$.

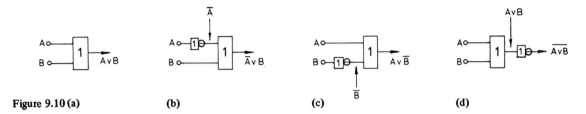

Figure 9.10 (a)   (b)   (c)   (d)

In a similar manner, *Figure 9.10* shows four possible combinations for the logical OR in conjunction with NOT. By combinations of this kind, a great variety of expressions and conditions essential to computer design in particular can be evolved. Some of these are more important than others and we shall discuss two of these in a little more detail.

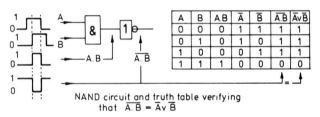

Figure 9.11

1. NOT AND circuit. This circuit, known as the NAND element, is shown in *Figure 9.9(d)* and illustrated in more detail in *Figure 9.11*. When A.B is applied to the input of the NOT gate, the output is $\overline{A.B}$. Notice that the negation bar is drawn across both A and B. This means that the inverter negates A to $\bar{A}$, AND to OR, and B to $\bar{B}$. Therefore, as a glance at the accompanying truth table will verify,

$$\overline{A.B} = \bar{A} + \bar{B}$$

You may recognise this expression as one of De Morgan's rules which you verified (or should have) in Problem 19 of the previous section. The NAND gate is a very important element in logical circuit design.

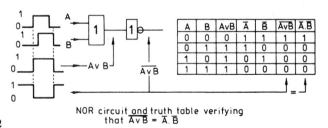

Figure 9.12

2. NOT OR circuit. This circuit, known as the NOR element is shown in *Figure 9.10(d)*, and illustrated in more detail in *Figure 9.12*. Like the NAND circuit, this has a number of inputs and one output.

102 Combinational logic gates

When there is a signal on one of the inputs or on any combination of the inputs, there is a NOT signal at the output. In symbols, $\overline{A + B}$ (notice again that the negation bar implies that NOT applies to the whole term) means that the inverter negates A to $\overline{A}$, the OR to AND, and B to $\overline{B}$. Again, referring to the truth table, we see

$$\overline{A + B} = \overline{A}.\overline{B}$$

This is the second of De Morgan's rules. Like the NAND gate, it is an important element in logical design.

Make sure you are familiar with the NAND and NOR symbols from *Table 9.1*.

Example (7). Devise a circuit whose output represents the function $F = A.B + \overline{A}.\overline{B}$, given two inputs A and B, and using only AND, OR and NOT elements.

We notice that the terms $A.B, \overline{A}.\overline{B}$ are linked by the OR symbol; these two terms will therefore be the output of an OR gate which has the inputs $A.B, \overline{A}.\overline{B}$. This situation is illustrated in *Figure 9.13(a)*. The $\overline{A}.\overline{B}$ term has to be the output of an AND gate having inputs $\overline{A}$ and $\overline{B}$; this brings us to diagram *(b)*. Similarly, the $A.B$ input term will be the output of another AND gate for which the inputs are respectively A and B: diagram *(c)*. To complete the picture, we can produce $\overline{A}$ and $\overline{B}$ signals by inversion of the A and B primary inputs. The final circuit then appears at *(d)*.

It is important to keep in mind that the solution given is not necessarily the simplest or most economical way of achieving the required expression.

Example (8). Devise a combination of AND, OR and NOT elements which will perform the function of an exclusive OR gate.

We have already noted that the exclusive OR gate gives an output only if A = 1 or B = 1 but not if both equal 1. If we look at the truth table for an exclusive OR gate, we see that the output we want is represented by the expression

$$F = \overline{A}.B + A.\overline{B}$$

A	B	F
0	0	0
0	1	1	→ $\overline{A}.B$
1	0	1	→ $A.\overline{B}$
1	1	0

Figure 9.14 shows the required combination. It is derived by reasoning in the same way as we did in the previous example. $\overline{A}.B$, $A.\overline{B}$ are connected by the OR symbol, hence these are the outputs from an OR gate having the separate inputs $\overline{A}.B$ and $A.\overline{B}$. These two terms are in turn connected by the AND symbol, so each is the output from AND gates having respective twin inputs A and $\overline{B}$, $\overline{A}$ and B. The negated signals are derived in turn from two NOT gates wired into the primary A and B input lines as shown.

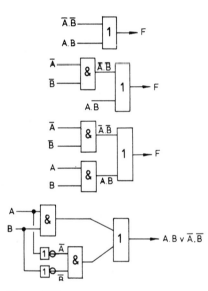

Figure 9.13

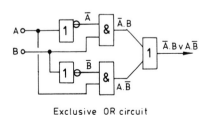

Exclusive OR circuit

Figure 9.14

DIODE-TRANSISTOR LOGIC

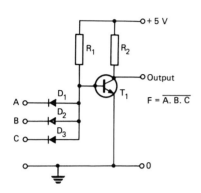

Figure 9.15

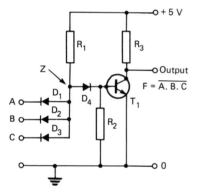

Figure 9.16

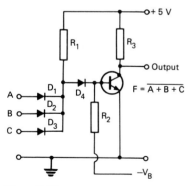

Figure 9.17

The logic gates available in modern integrated circuits are rather more sophisticated than the simple diode-resistor combinations we have discussed so far, and we now look at a few different families of logic systems to see how development has taken place.

The obvious step in possibly improving on the diode-resistor circuits is to follow the diodes with a transistor switch in a single system. This is a basic diode-transistor logic gate (DTL) which can perform either the AND plus NOT function or the OR plus NOT function, that is, either the NAND or the NOR function. A DTL NAND gate is shown in *Figure 9.15*. The value of resistor R_1 is now chosen to suit the switching requirements of the transistor T_1. When all inputs A, B and C are high, the diodes are switched off and the transistor is held in saturation by the positive bias on its base. The output at its collector is then low. If one or more inputs are taken low, the relevant diode (or diodes) switch on and the base of the transistor goes low. If this base voltage falls below about 0.6 V, the transistor will turn off and its output will go high. We thus have the NAND function where the output is low only if *all* the inputs are high.

There is, however, a difficulty with this circuit as it stands. If silicon diodes are used, as they invariably are, the transistor will either not be turned completely off or it will only *just* be turned off, since the forward voltage drop of the diode will also be in the region of 0.6 V. This situation is much too risky for comfort. Some improvement will be possible by using germanium diodes where the forward drop will be about 0.2 to 0.3 V, but a better arrangement is to use a so-called level-shifting diode as shown in *Figure 9.16*. As before, when all the inputs are high, all the diodes on the inputs are cut off and the current flowing through R_1 and D_4 (which is forward biased) makes the base of the transistor sufficiently positive to ensure saturation, the voltage level at point Z being $2 \times 0.6 = 1.2$ V, not 0.6 V as it was before D_4 was included. It is quite usual to have two diodes in series for D_4, so raising the potential at point Z to about 1.8 V.

When any one or more of the inputs go low, the voltage at point Z becomes 0.6 V but this time the presence of D_4 prevents sufficient current flowing into the base of the transistor to permit it to conduct, hence it switches off and the output goes high.

A DTL NOR gate is shown in *Figure 9.17* and you should have no difficulty in deducing how it operates. Look at it as an OR gate followed by a NOT gate. The transistor is switched hard on when one or more inputs are taken to logical 1, that is, the output is high only when *all* inputs are low.

TRANSISTOR-TRANSISTOR LOGIC

In integrated circuit technology it is just as simple to make a special transistor as it is a diode, and the next step is to replace the separate diodes used in the DTL circuits with a multiple emitter transistor as shown in *Figure 9.18*. Not only does this transistor replace the separate

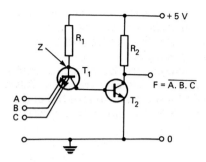

Figure 9.18

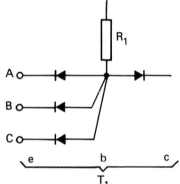

Figure 9.19

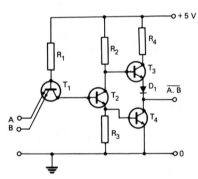

Figure 9.20

diodes, it replaces the level shifting diode D_4 of *Figure 9.16* as well by the presence of the base-collector junction. A clearer understanding of the way the special transistor does its job can be gained by interpreting the circuit in the form shown in *Figure 9.19* where its similarity to *Figure 9.16* will be apparent. Circuits of this sort are known as transistor–transistor logic or TTL.

When all input levels are high, none of the base-emitter junctions of T_1 is forward biased. The base-collector junction is, however, forward biased and current flows through this diode into the base of T_2 saturating it and causing the output to go low. So that T_2 can be kept saturated, it is necessary for point Z to be at about 1.2 V as we noted for the DTL circuit above. If one or more of the emitters are now taken low, those base-emitter junctions will switch on and the voltage at point Z will fall to about 0.6 V. As for the DTL circuit of *Figure 9.16*, this level will be insufficient to permit current to flow into the base of T_2 which will switch off and allow the output to go high.

Now although this circuit is an advance on the discrete diode systems, there is still a problem when it becomes necessary to connect the output of the gate to following gates. The ability of a logic gate to feed its output into a number of following gates without deterioration or erratic performance is known as its *fan-out* value. For a TTL form of gate element, this is typically 10, but to achieve this figure it is necessary to modify the output arrangements so far described in the circuits of *Figures 9.15* to *9.18*. We will look into problems associated with fan-out shortly; in the meanwhile, we look at a common form of output circuit found in most TTL integrated circuit gates.

The single transistor output stage so far considered is replaced by two transistors in what is known as a *totem pole* circuit, and this is illustrated in *Figure 9.20*. This diagram shows a complete integrated NAND gate which includes the multiple emitter input arrangement already discussed. For simplicity only two inputs, A and B, are considered. When A and B are both high, neither base-emitter junction of T_1 is forward biased but the base-collector junction is, so that current flows through it via R_1 into the base of T_2, switching it hard on together with T_4. With T_2 on, the base-emitter potential of T_3 is too low for forward bias and T_3 is off. So if all inputs are high, T_2 and T_4 are switched on, T_3 is off, and the output is low. The diode D_1 is there to ensure that its forward voltage drop of 0.6 V will keep T_3 off when T_4 is on.

If now one or both of the inputs fall low, the emitter(s) of T_1 is effectively at zero potential and its collector voltage falls, turning off T_2. The collector of T_2 then rises towards V_{CC} and the emitter falls to zero volts. T_3 then conducts (base high) while T_4 switches off (base low). The output then goes high. Notice from all this that when the output is high, T_3 behaves as a grounded collector (or emitter follower) stage which drives current into whatever load is connected to the output. This is known as current 'sourcing'. When the output is low, current flows from the load into T_4 which is saturated and has a very low resistance. This is known as current 'sinking'. So the totem pole arrangement presents a low resistance output in both logic states: either via T_4 when the output is low or via T_3 when the output is high. This ability to source and sink relatively large currents from and into low resistance paths gives the totem pole circuit a great advantage over the single transistor output stage. We will go into this in the next section.

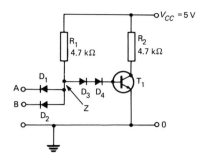

Figure 9.21

Example (9). For the DTL NAND gate of *Figure 9.21*, the output transistor has $a_E = 50$ and a V_{CE} of 0.2 V when saturated. If input A is at 5 V and input B is at 0.4 V, what will be the output voltage at the collector of T_1? If input B now rises to 2 V, what will be the output voltage?

One output (A) is high, the other (B) is low. Since the state of a NAND gate is determined by the *lowest* input and B = 0.4 V, the potential at point Z (V_Z) cannot exceed 0.4 + 0.6 = 1 V. If current flows through the level-shifting diodes D_3 and D_4, the base-emitter voltage of T_1 is

$$V_{BE} = V_Z - 2(0.6) = 1 - 1.4 = -0.4 \text{ V}$$

This condition ensures that T_1 is completely cut off and the collector voltage stands at $V_{CC} = 5$ V.

When input *B* rises to 2 V, V_Z cannot exceed 2 + 0.6 = 2.6 V, but for T_1 to conduct, V_Z must exceed 3(0.6) = 1.8 V, hence base current will flow of a value

$$I_B = \frac{V_{CC} - 1.8}{R_1} = \frac{5 - 1.8}{4700} \text{ A} = 0.68 \text{ mA}$$

As the collector load R_2 is 4700 Ω, the greatest possible collector current would be

$$I_{C(\max)} = \frac{V_{CC}}{R_2} = \frac{5}{4700} \text{ A} = 1.06 \text{ mA}$$

and a base current of $I_C/a_E = \frac{1.06}{50} = 0.021$ mA ;

would be sufficient to produce this.

As the actual base current is 0.68 mA, the transistor is driven hard into saturation and the output voltage is now $V_{CE(\text{sat})} = 0.2$ V.

This example shows that the input logic levels do not have to be equal or indeed up to the 5 V level for the state of the gate to be determined. What we refer to as 'high' or 'low' must clearly cover a range of voltage levels for a particular gate. We shall see that a 2 V level is right at the foot of the accepted range for which the term 'high' is applied.

(10) Deduce that a NAND gate with all its inputs joined together behaves as a NOT gate.

(11) Connect together two NOT gates and one NOR gate to perform the function of an AND gate.

(12) Use two NOR gates and an AND gate to make an exclusive OR gate.

COMMERCIAL INTEGRATED LOGIC

We have now covered all the basic logic gates and a few of the possible combinations. Throughout we have assumed that these circuits are made up of discrete component parts, resistors, diodes and transistors, and indeed, gates can be made up in this way for many experimental investigations. For very complicated logical systems, however, even going no further than a simple hand-held calculator where several hundred gates may be necessary, such discrete component assemblies would become large and unwieldy, as well as consuming considerable power. Logic families have, as a consequence, appeared over the past twenty years or so in the form of integrated circuits and in a great variety of gate arrangements.

The earliest family was the 7400 series which used transistor-transistor logic and of which *Figure 9.20* is an example. This family is available to the present day at very low cost, each integrated package containing a number of identical gates. In the intervening period of evolution when a greater number of gates and functions were being called for, the basic TTL form was not wholly suitable for integration because the chip area necessary for each gate was too great and the heat dissipation excessive. The problem has been overcome by the introduction of large scale integration (LSI) utilising CMOS technology. Such circuits are found in the 4000 series, though this range in turn is being superseded by the 74HC series. We concentrate at this stage of the course on TTL and the 7400 series.

The 7400 Series

TTL gates are found in everyday DIL (dual-in-line) packages having the general appearance shown in *Figure 9.22*, where pin number 1 is marked with a dot or indentation. The packages have usually 14 or 16 pin-outs and each contains a number of gates, the nominal 5 V operating voltage (V_{CC}) supply pins being common to all of them. The family is identified by a number of letters in the form - - 74 - - or - - 75LS - -. The LS series offers a superior performance over the standard range in respect principally of power consumption, but the internal gate arrangements are otherwise identical. The first letters are usually a code for the name of the manufacturer.

Table 9.2 shows five popular 7400 integrated circuits, together with the pin connections. The basic gate of the series is the 7400 itself which is a quad two-input NAND gate, each unit containing the circuit arrangement shown earlier in *Figure 9.20*. The 7402 is a quad two-input NOR gate, while the 7404 contains six inverters or NOT gates. The 7408 is a quad two-input AND gate and the 7486 is a quad two-input

Figure 9.22

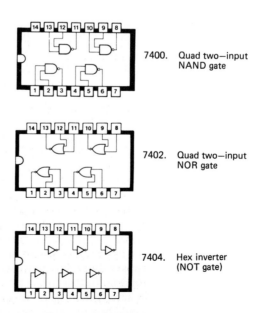

7400. Quad two-input NAND gate

7402. Quad two-input NOR gate

7404. Hex inverter (NOT gate)

Combinational logic gates 107

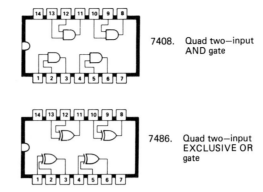

7408. Quad two-input AND gate

7486. Quad two-input EXCLUSIVE OR gate

Table 9.2

EXCLUSIVE OR gate. In a circuit system, one or all of the gates contained in the chip may be used, the V_{CC} supply being applied between pins 14 (positive) and 7 (negative ground). All diodes, transistors and resistors are formed on the silicon chip as a compact integrated assembly, and the cost is little more in many cases than the price of a single discrete transistor.

(13) TTL circuits are preferred to DTL because
  (a) they have a smaller fan-out;
  (b) they operate from a 5 V supply;
  (c) the multiple emitter transistor is easier to integrate than a number of diodes;
  (d) they are available in an extensive I.C family.
(14) The fan-out of a TTL gate is determined by
  (a) the output resistance of the gate;
  (b) the load current of the output;
  (c) the applied V_{CC} level;
  (d) the actual logic levels employed.

TTL Characteristics Manufacturers issue data sheets for each of their logic gates and a number of important operating parameters must be appreciated before any such gates are incorporated into a circuit system. You should obtain a few of these sheets and study them carefully. Here we shall consider some of their most important points.

1. *Operating voltage.* The 7400 series TTL gates operate from a nominal V_{CC} supply of 5 V, but there are limits to the variation which can be permitted from this figure. V_{CC} has an *absolute* maximum rating of 7 V, and if a variable power supply unit is being used in any experiment with TTL gates, great care must be taken that it is not set to a voltage greater than 7 V at switch-on or at any time subsequently. Any supply over this top limit will almost certainly destroy the gate. Using batteries is generally a safe way to avoid this possibility of damage; a 4.5 V battery, which when new will provide a terminal p.d. of some 4.75 V, is ideal. A 6 V battery may be used when nothing else is available, though the manufacturer's *operating* limit of V_{CC} is 4.75 to 5.25 V. Where it is important to keep the supply accurately at 5 V is in the matter of the speed of response of the gate. As V_{CC} is reduced, the time taken for a logical change to pass through the system increases. This may not be very important in simple bench experiments, but

circuits such as high-frequency counters and dividers would not operate to their full specifications under low V_{CC} conditions.

2. *Logic levels and noise margin.* Data sheets provide us with the voltage levels that a TTL gate will accept as a legitimate 1, or high, and the voltage levels it will accept as a logic 0, or low. We have, for convenience and simplicity, so far taken +5.0 V to be high and 0 V to be low, but as in the V_{CC} supply, there are limits to what the input levels may be in order to assure trouble-free operation. Generally, the minimum logic 1 level, V_{IHMIN}, is taken to be 2.0 V, and the gate is guaranteed to recognise an input of 2.0 V or more as a high and accept it as such. The maximum logic 0 level, V_{ILMAX}, is taken to be 0.8 V and any input signal equal to or less than 0.8 V is accepted by the gate as a legal low. *Figure 9.23* illustrates these level conditions. Notice that there is a region between 0.8 V and 2.0 V which comes into neither the high nor the low level category; this region is indeterminate and input signals must not settle at levels within this band.

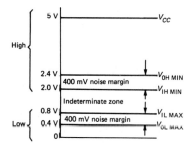

Figure 9.23

Two other levels marked in *Figure 9.23* need our attention. These relate to the *output* voltage levels obtained from a gate. For the 7400 series, an output logic 1 is guaranteed to be equal to or greater than a minimum level V_{OHMIN} of 2.4 V. In actual practice, output highs are usually above 3.0 V. Make a note that the *minimum* output high (2.4 V) is 0.4 V or 400 mV higher than the minimum level required for an *input* high on a following gate. This 400 mV gap is known as the *noise margin*. Noise margin is a measure of the ability of a gate to reject a small noise impulse at its input and prevent it from erroneously changing state. A noise impulse has to be greater in magnitude than 400 mV before it can force the input to a following gate to go below the 2.0 V input high minimum. Similarly, a noise margin of 400 mV exists between the maximum logic 0 output level of 0.4 V and the maximum logic 0 input level of 0.8 V. Noise margins can be affected, and nullified, by excessive loading on a particular gate.

3. *Sink and source currents.* These terms were mentioned in passing earlier on; we now examine them in more detail. On the manufacturer's data sheets you will find output currents marked for logic 1 and logic 0 output voltage conditions. These are usually stated as 400 µV for logic 1 and 16 mA (often −16 mA) for logic 0 outputs, but we can divide these figures by a factor of 10 since they are given for a fan-out of 10, that is, when the gate output in question is connected to the inputs of ten others.

We consider the case, then, of the output of one gate connected to the input of another, taking as our example the NAND gates of *Figure 9.20*. *Figure 9.24* shows the situation. Relate the 'internal' components shown in the gates to the totem pole output circuit in the one on the left, and part of the multi-emitter input circuit to the one on the right. Now in diagram *(a)* we assume that the output of gate G_1 is low; it then creates a current path (through the switched-on lower transistor of the totem pole) for the V_{CC} supply of gate G_2 to earth as indicated by the broken line. Gate G_1 is then *sinking* current as it pulls the input to gate G_2 low.

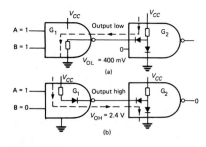

Figure 9.24

The manufacturer's data states that the maximum logic 0 input current for a 7400 TTL gate (I_{LMAX}) is −1.6 mA. The negative sign is simply a convention that indicates that the current (conventional) is flowing *out* of the relative input terminal; a positive sign indicates that the current is flowing *into* that terminal. So the figure of 1.6 mA means that for a TTL input to be pulled low, the output connected to it *may* have to sink a current of that magnitude to earth.

Figure 9.24(b) shows the situation when the output of gate G_1 goes high. A current now flows (conventionally) from the V_{CC} supply of gate G_1 (through the switched-on upper transistor of the totem pole) into the input of gate G_2, again indicated by the broken line. This current flows through gate G_2 to earth. So gate G_1 is said to *source* current to gate G_2 in order to pull the input to G_2 high. The maximum logic 1 input for a TTL gate is 40 μA, so gate G_1 *may* have to provide a current of that magnitude to earth.

We have noted that fan-out is a measure of the greatest number of gate inputs which may be connected to a single gate output without preventing the output from reaching the permissible high and low voltage levels. The fan-out for TTL gates is 10; the total available source current is guaranteed to be at least $(10 \times 40 \mu A) = 400 \mu A$, and the guaranteed sink current is $(10 \times 1.6 mA) = 16$ mA. You should now be able to appreciate the usefulness of the totem pole output circuit relative to the single transistor of the RTL circuits.

4. *Open-collector outputs*. In spite of what we have just said about the totem pole, the configuration has a disadvantage when two or more outputs from separate gates are to be connected together. You should be able to surmise for yourself what is likely to happen if you connect together the output terminals of two (or more) totem pole circuits. If, glancing back to *Figure 9.20*, the T_4 of one gate happens to be switched on when the T_3 of the other gate is switched on, a low resistance path to earth is provided through them and the relevant diode, so effectively short-circuits one of the outputs. Resistor R_4 helps to reduce the current, but this resistor is not always included; make sure, then, that you never short the output of a TTL gate, either directly or by an attempt to connect two gates directly together.

One way out of the difficulty, of course, is to connect the outputs from, say, three NAND gates to a fourth gate as shown in *Figure 9.25*, this being followed by a NOT gate. The output is then clearly $F = \overline{AB.CD.EF}$. However, if the circuit of the totem pole is modified by omitting the upper transistor and its resistor (see *Figure 9.26*) we have what is known as an 'open-collector' output. When several outputs are now joined together, a *single* resistor can take the place of the original individual resistors; this is known as a pull-up resistor. When the transistor T_4 is off, this resistor brings the output to a high. The 7400 series has a family of gates with open-collector outputs: the 7401 and 7403, for example, are both quad two-input open-collector NAND gates, the 7405 is an open hex inverter and the 7409 is a quad two-input open-collector AND gate.

The open collector output system tends to be noisier than the standard totem pole configuration and slower to respond to a change of state, and these are disadvantages. For this reason, the open-collector has given way in many cases to what is known as the *three-state* output logic where the output has the normal TTL low and TTL high, but also has a floating output state created by turning off both output transistors with a separately applied signal. This technology will be covered in later years of the course.

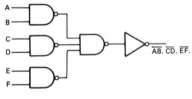

Figure 9.25

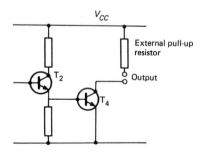

Figure 9.26

The 4000 Series

Most of your work in logic systems at this stage of the course will be done with the 7400 TTL series of integrated circuits. It is necessary, however, to be aware of the 4000 series of logic gates which make use of the CMOS-based (or *complementary metal oxide semiconductor*) technology. These gates make use of the field effect transistor which gives a number of advantages over the TTL series that use the bipolar transistor. These advantages include a very small current consumption, a wider range of operating voltage, typically from 3 V up to 15 V, and a high fan-out capability brought about by the very high input resistance of CMOS gates.

Like the TTL series, the CMOS series is numbered from a basic 4000 upwards. The 4001, for instance, is a quad two-input NOR gate, the 4011 a quad two-input NAND gate and the 4049 a hex inverting gate.

The basic arrangement for a 4000 series inverter is shown in *Figure 9.27(a)*, with a NAND gate in *(b)*. The inverter is built up from one N-channel and one P-channel enhancement MOSFET as described in an earlier chapter. When the input is low (logic 0), the N-MOSFET is cut off and the P-MOSFET is switched on. Thus there is a low resistance path from the supply rail V_{DD} to the output; hence the output is high and closely equal to V_{DD}. When the input goes high (logic 1), the transistors change state and the output is taken low through the switched-on N-MOSFET. The circuit thus behaves as an inverter or NOT gate.

This gate can be extended into a NAND gate by the addition of two more devices, *Figure 9.27(b)*. It is left as an exercise for you to deduce that when the inputs A and B are both high the output is low, and when either A or B or both are low the output is high.

The logic levels for the 4000 series are shown in *Figure 9.28*. They are given as percentages of the applied V_{DD} which, as we have noted, can be anything from 3 V to 15 V. Notice that the noise margins are here one-third of the level of V_{DD}.

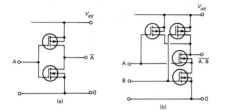

Figure 9.27

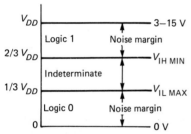

Figure 9.28

GATE ECONOMY

It is possible to make up any logical circuit combination to perform any logical function using only NAND or NOR units, and it is common practice for circuits to be designed in this way. The obvious advantages are (a) the economy obtained in bulk buying of gates, (b) the ease of replacement, hence ease of maintenance, and (c) the fact that logic levels remain nearly constant throughout the system.

We have noted earlier that a NAND gate with all its inputs connected together acts as a NOT gate. *Figure 9.29(a)* and *(b)* show respectively how NAND gates can be used to perform the AND and OR functions. The AND gate in diagram *(a)* uses two NAND elements, the first used as a NAND and the second as a NOT. The output is then the same as that of a single AND gate.

In diagram *(b)* three NAND elements are used to perform the function of an OR gate. The first two NANDs are employed as NOTs, followed by a normal NAND; the output is then a double-negated product which is equivalent, by De Morgan's theorem, to the OR gate A + B.

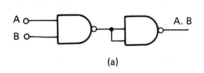

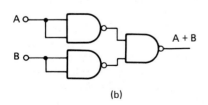

Figure 9.29

> (15) Using NAND gates only, make up circuits performing the functions of (a) NOR gate, (b) EX-OR gate.

It is an interesting point that though NAND and NOR gates may be used to make any other type of gate (and are often known as 'universal gates' for this reason), AND and OR gates can never be assembled in any manner to make NAND, NOR or NOT relationships.

> *Example (16)*. Show that the circuit of *Figure 9.30*, which uses four different gate elements, can be replaced by a single NAND gate.
>
> This kind of question can be solved by the use of truth tables or by the application of logical algebra. We will illustrate the use of logical algebra in this case. On the diagram the various intermediate logical conditions have been indicated and the final output is $F = \overline{D} + \overline{C(A.B)}$. This does not look like the output of a single NAND gate, but if we apply our algebraic rules we get:
>
> $$F = \overline{D} + \overline{C(A.B)} = \overline{D} + \overline{C} + \overline{(A.B)} \text{ by De Morgan}$$
>
> Applying De Morgan a second time, we have
>
> $$\overline{D} + \overline{C} + \overline{(A.B)} = \overline{D.C} + \overline{A.B} = \overline{A.B.C.D}$$
>
> which is the output of a four-input NAND gate. Why use four gates when one will do?

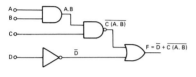

Figure 9.30

SOME HINTS ABOUT USING TTL AND CMOS

We conclude with a few important hints about the way TTL and CMOS gates should be used and handled in bench experiments. CMOS devices are particularly sensitive to mishandling and careless work can easily lead to the destruction of the integrated package.

For TTL, any unused inputs should be either connected up to the V_{CC} line by way of a 1 kΩ resistor or tied directly to the earth (zero) line. Remember, also, that the *maximum* V_{CC} permitted is 7 V; try to keep the V_{CC} line as close to 5 V as you can.

The outputs of TTL gates must not be connected together unless you are dealing with open-collector devices. Any unused outputs should be left floating.

When experimenting with TTL gate circuits, particularly on protoboards, plug-in systems, S-Decs or Veroboard, always shunt the V_{CC} supply points (usually pins 7 and 14 on TTL packages) with a ceramic 0.1 μF capacitor, keeping this as close to the pins themselves as possible. Also, keep all the interconnecting leads as short as possible. Anything longer than about 25 cm can lead to instability problems.

CMOS handling is more critical than TTL. Like TTL, all unused CMOS inputs should be connected to V_{CC} or earth, but the inputs can be damaged by discharge of static electricity built up on the fingers or particularly nylon shirt cuffs. Always make sure that the soldering iron you are using has its case securely earthed. Keep all CMOS devices wrapped in foil (aluminium kitchen foil will do) or embedded on conducting plastic until they are ready for use. Do not mix up TTL and CMOS packages in one circuit system. (They *can* be intermixed, but not at this stage of the course!) Finally, keep all leads as short as you can, and fit bypass capacitors across the supply pins of each package.

PROBLEMS FOR SECTION 9

(17) Complete the following statements:
  (a) When logical 0 is at the input of an inverter, the output signal is logical . . . .
  (b) The output $A \oplus B$ is provided by a(an) . . . . gate.
  (c) A NOR circuit includes both . . . . and . . . . elements.
  (d) All logic functions can be constructed using only . . . . . or . . . . gates.
  (e) NAND followed by NOT is a(an) . . . . gate.

(18) Draw the logic symbols for the AND, OR, NAND, NOR and NOT gates using (a) BS and (b) ANSI symbols. What are the equivalent symbols for the EX-OR gate?

(19) Write down the output expressions for the logic circuits shown in *Figure 9.31*.

(20) For the logic system shown in *Figure 9.32*, deduce that the output will be high if and only if A = B and C = D.

(21) Deduce that the circuit of *Figure 9.33* will give a high output if an odd number of inputs is high and a low output if an even number of inputs is high.

(22) A student built up the circuit shown in *Figure 9.34(a)* as a design for an AND gate, using two p-n-p transistors. Explain why this circuit failed to work. He then modified the circuit to that shown in *Figure 9.34(b)*. Did it work this time? Whatever you decide, explain your reasoning.

(23) In the circuit of *Figure 9.35* the input signal levels are either 0 V or 10 V. Answer the following:
  (a) Under what condition(s) will the potential at X be 10 V?
  (b) What then will be the potential at Y?
  (c) What logic function does this circuit perform?

(24) A logic circuit has inputs A and B to an OR gate whose output is AND-gated with input C to give an output X.

A second circuit has two AND gates with separate inputs A and B and which share input C. The outputs of these gates enter a two-input OR gate to give an output Y.

Draw these circuits and show by reasoning and a truth table that output X is identical with output Y.

(25) How would you recognise pin 1 on a dual-in-line integrated package? How are the remaining pins numbered relative to pin 1?

(26) Define noise margin. What are the noise margins for TTL gates?

(27) Sketch the circuits, using any gates you like, which satisfy the output expressions:
  (a) $F = AB + BC$ (b) $(A+B)(C+D)$ (c) $B + \overline{A.C}$

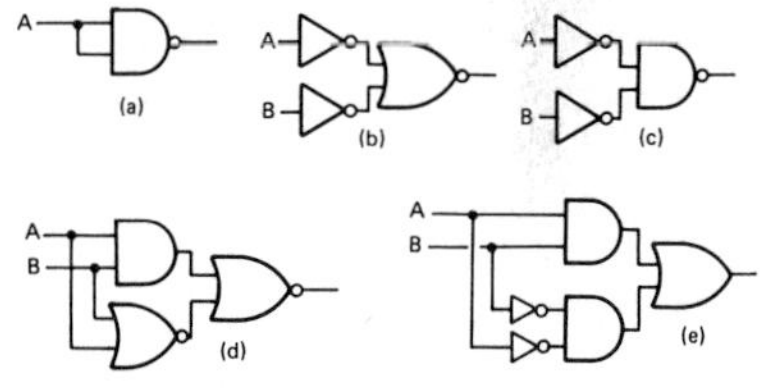

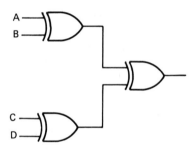

Figure 9.31

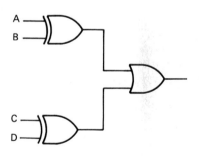

Figure 9.32

Figure 9.33

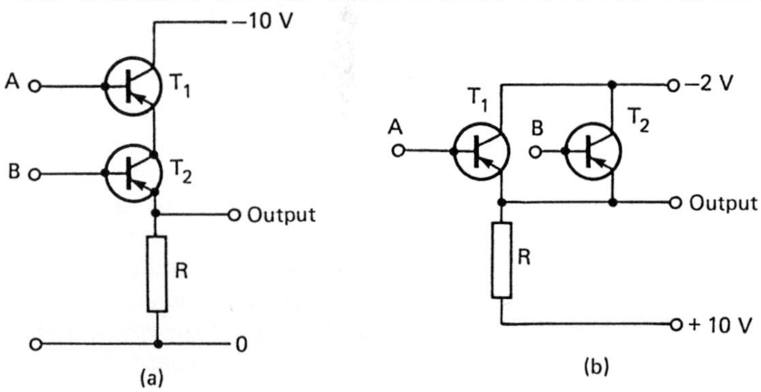

Figure 9.34

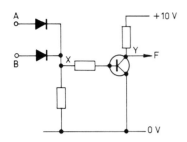

Figure 9.35

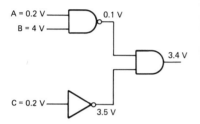

Figure 9.36

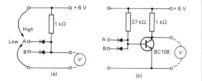

Figure 9.37

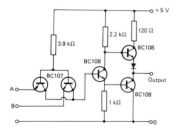

Figure 9.38

Combinational logic gates 113

(28) Now try designing the circuits you obtained for problem 23 using only NAND or NOR gates.

(29) The logic assembly shown in *Figure 9.36* exhibits a fault. A student takes voltage measurements at all the terminal points and records them as shown. Which part(s) of the circuit is giving trouble?

(30) A machine operator controls red and green indicator lights by four switches A, B, C and D. The sequence of operation is as follows:

(i) Red light is ON when switch A is ON and switch B is OFF or switch C is ON.

(ii) Green light is ON when switches A and B are ON and switches C and D are OFF.

Write down logical expressions representing conditions (i) and (ii), and sketch logical circuits satisfying these expressions.

(31) Use NOR gates to form a two-input AND gate and check your solution by means of a truth table. Now do the same thing using only NAND gates to produce an OR gate.

(32) Explain why it is preferable to use only one type of gate in a logic system.

The following questions refer to an experimental circuit set-up.

(33) Build the circuit shown in *Figure 9.37(a)*, connecting a high resistance voltmeter (an Avometer is suitable) set to the 10 V range where shown. The diodes should be silicon types such as the 4148.

Connect inputs A and B in turn to the high line (6 V rail) and note the voltmeter reading. Repeat this by connecting the inputs A and B in turn to the zero line. From the levels recorded by the voltmeter, what kind of gate is this?

Now add a transistor stage as shown in *Figure 9.37(b)*. Repeat the above input conditions and record the output levels again. What kind of gate is it this time?

(34) Simulate a NAND gate using discrete components as illustrated in *Figure 9.38*. The two input transistors (which are n-p-n types) simulate the multiple emitter transistor found in integrated packages. Using the voltmeter to indicate the output logic level, and connecting A and B to high (V_{CC}) or low (0 V) as appropriate, complete the truth table below.

A	B	Output
0	0	
0	1	
1	0	
1	1	

You may, if you wish, use a logic probe as indicator (if you have one available). Or a digital voltmeter will do.

10 Karnaugh mapping

Aims: At the end of this Unit section you should be able to:
Understand the structure of a Karnaugh map.
Construct a map either from logical expressions or from truth tables.
Use the map to minimise logical expressions.

We have seen how we can set up logical expressions from truth tables or from a study of a particular arrangement of logic gates. We have also seen that, like its counterpart in ordinary algebra, a logical expression can often be simplified to give a shorter, neater form of the equation. In creating a circuit to perform a particular logical function, this fact can be used by an engineer to produce an optimum design. In most cases this will be a design objective involving the least number of gates (if possible, identical gates), together with the criterion of minimum cost. Reducing a logical expression for such a purpose is known as *minimisation*.

If a truth table is available or if the logical function can be written in the form of the *sum-of-product* terms, e.g. $\overline{A}BC + A\overline{B}C + A\overline{C}$, the engineer can go directly to the minimal expression by a mapping technique named after its originator, Maurice Karnaugh.

Karnaugh mapping (or K-mapping) is a graphical way of displaying logical expressions or switching function. We shall here be looking only at the basic use and elementary applications of this system, but in its full development it provides a powerful method for the minimisation of functions as well as for the proof of theorems and the translation of complicated logical circuits into other forms using different and possibly cheaper gates.

THE MAP STRUCTURE

In a function having n variables, these are 2^n ways of combining the variables. So, if we construct a two-dimensional diagram having a square (or cell) for each possible combination of the input variables, we shall have an array made up of $2^2 = 4$, $2^3 = 8$, $2^4 = 16$ etc., squares or cells, suitable respectively for the representation of logical functions having 2, 3 or 4 variables. The case of a single variable ($n = 1$) is of no particular interest.

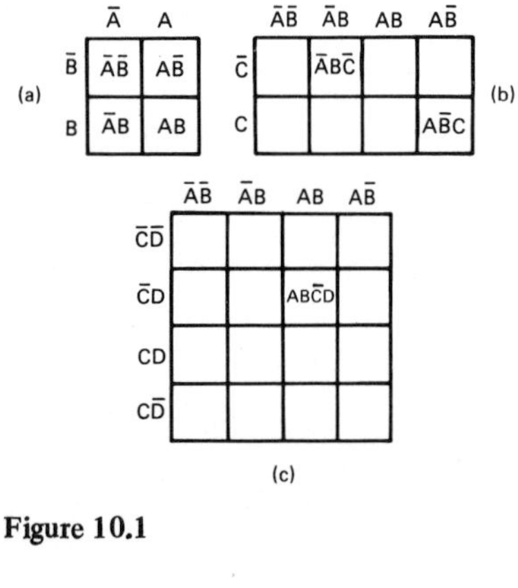

The diagrams shown in *Figure 10.1(a), (b)* and *(c)* are of maps having 4, 8 and 16 cells respectively. Each cell within the maps is at the intersection of an externally indicated column and row. In this way, each cell has a unique position (or address) which is defined completely by the labels attached to the columns or rows which intersect at that particular cell. In diagram *(a)*, the four cells have been marked with their addresses. Notice that the address of any cell *differs by no more than one symbol from the address of any adjacent cell*. In this respect, diagonal cells are not considered adjacent. We shall come back to this point shortly.

The second map shown at *(b)* has two of its cells marked with their addresses; these are the cells $A\overline{B}C$ and $\overline{A}B\overline{C}$ or (A and not-B and C) and

Figure 10.1

(not-A and B and not-C). In diagram *(c)* the marked cell has the address $AB\bar{C}D$ (A and B and not-C and D). Sketch these diagrams on a sheet of rough paper and fill in for yourself all the cell addresses as was done for diagram *(a)*. When you have done this, check carefully and notice that the addresses conform to the note made above: no cell differs by more than one symbol from the address of any adjacent cell. Don't go further until you are quite certain you understand these basic points about the Karnaugh map.

A point of particular importance is the method of labelling the variables around the perimeter of the map. The arrangement used in *Figure 10.1* is not the only one possible, but it is possibly the most convenient and the most easily remembered. When the map is actually used, it is often more customary to insert the logic levels 0 and 1 around the perimeter with the variables indicated at the top left-hand corner, as shown in *Figure 10.2(a)* and *(b)*. This makes it easy for mapping from a truth table because each cell in the map corresponds to a row in the truth table. It is very important to notice the sequence of 0s and 1s along the sides: 00, 01, 11 and 10, *not* 00, 01, 10 and 11.

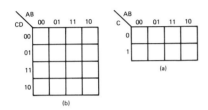

Figure 10.2

FILLING IN THE MAP

Suppose that we have an expression of, say, three variables A, B and C, given in the form of a sum-of-product terms such as those represented by single map cells. For example, suppose

$$f(A,B,C) = AB\bar{C} + A\bar{B}\bar{C} + A\bar{B}C$$

This function will be mapped as shown in *Figure 10.3*, where each of the three cells designated by the function is marked with a logical 1. The other cells are left blank. The total diagram area covered by 1s now represents the overall function. It should be evident that if a map has *all* its cells occupied by 1s, the function corresponding to this must display unity transmission for all possible combinations of all its variables, that is, the relevant logical circuit would give a 1 output for any possible combination of input. Hence such a circuit would behave as a permanently closed circuit. A map having no cells occupied would likewise represent a permanently open circuit, with 0 output for any input combination. Both circuits of this sort would simply be an expensive way of achieving nothing!

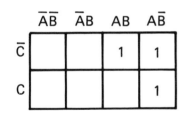

Figure 10.3

The concept of these two extremes, however, has its uses. Since the cells marked with 1s represent the overall function f(A,B,C) the remaining (or zero) areas must represent the sum of those terms which produce zero transmission, hence it must represent the complement $\bar{f}(A,B,C)$ of the original function. From *Figure 10.3*, therefore we deduce that

$$\bar{f}(A,B,C) = \bar{A}\,\bar{B}\bar{C} + \bar{A}\,\bar{B}C + \bar{A}B\bar{C} + \bar{A}BC + ABC$$

(1) How many cells are required to map the following function?

$$f(A,B,C) = AB\bar{C} + \bar{A}BC + \bar{A}\,\bar{B}C$$

Draw the map and mark the cells designated by the function with 1s. Write down the complement of this function.

116 Karnaugh mapping

TRUTH TABLE TO K-MAP — A Karnaugh map can be constructed directly from a truth table without the necessity of having to write out the logical equation. An example will illustrate the method.

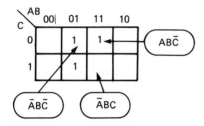

Figure 10.4

Example (2). A truth table for a three-variable logic circuit is shown below.

A	B	C	F
0	0	0	0
0	0	1	0
0	1	0	1
0	1	1	1
1	0	0	0
1	0	1	0
1	1	0	1
1	1	1	0

$F = 1$

Use this table to map the logical equation it represents.

We are going to map from the truth table, but as an exercise we first deduce the logical equation. This is done by taking out those rows of the table where the output F is logical 1. From the indicated rows we get

$$F = \overline{A}B\overline{C} + \overline{A}BC + AB\overline{C}$$

This could be mapped in the way already described on an eight-cell diagram. Using the map labelling shown in *Figure 10.2* earlier, however, we can plot the relevant rows from the truth table directly on to the map. This has been done in *Figure 10.4*, where a 1 is placed in each cell which has an address corresponding to that row in the table for which the output F = 1. Study this carefully; the product terms have simply been added for clarity and would not normally be shown. We will use this form of labelling from now onwards.

(3) Draw up the truth table for the logic circuit shown in *Figure 10.5*. Use an eight-cell diagram to draw the Karnaugh mapping for this table.

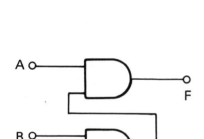

Figure 10.5

USING THE K-MAP — We now know what a K-map is and how we can map a function either from its truth table or from its logical equation. We want now to find out how we can use the map to give us useful results, or how it can help us in minimising logical expressions. Start off by looking at *Figure 10.6*. In this circuit, the intermediate outputs have been noted on the diagram in the usual way and the final expression for the output is $F = \overline{AB} + \overline{A} + B$. Notice that the circuit uses a NAND, a NOR, an inverter and one OR gate, a total of four gates. If we draw up the truth table, as has been done in the figure, we get an interesting result. Do you recognise F in the form it is now expressed? You should; we get an output 1 *except* when inputs A and B are both 1s. But this is the output we get from a single two-input NAND gate. Hence the circuit shown is simply doing the work of a single NAND gate; the other three gates, the NOR, NOT and AND gates are redundant to the operation. This illustrates the importance of minimisation; it enables complex circuits to be reduced and simplified. A Karnaugh map gives another

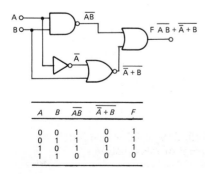

Figure 10.6

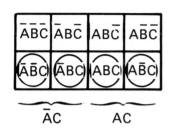

Figure 10.7

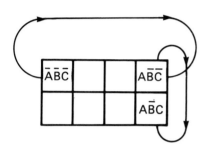

Figure 10.8

way of doing this, aside from truth tables and logical algebra.

We return now to the note we made about adjacent cells and the fact that two cells are adjacent in a map if their product term representations differ in one only of their variables. For example, the two cells of *Figure 10.7*, $\overline{A}\overline{B}C$ and $\overline{A}BC$, where the single difference is in the B variable, are adjacent. So are the cells ABC and $A\overline{B}C$, where the difference is again in the B variable. Further, to go from one such cell to the other involves crossing one boundary only, in the above instance the B boundary in both cases.

A look at the map shows us that in this sense the outer boundaries of the map are only *single* boundaries so that to go from, say, $\overline{A}\overline{B}C$ to $A\overline{B}C$ (which are adjacent according to our definition) involves not only crossing the A boundary but also entails shifting out of the map entirely, as shown in *Figure 10.8*. Both of those outer boundaries are equivalent. The same is true if we go from, say $A\overline{B}\overline{C}$ to $A\overline{B}C$, which is also depicted, this time crossing the C boundary. The outer limits of the map may therefore be regarded as being connected together, top to bottom and side to side, or, if you have an agile imagination, you might visualise it as being wrapped around a cylinder in both the horizontal and vertical directions. As already mentioned, diagonal cells are *not* adjacent; they always differ in more than one variable.

A pair of adjacent terms (cells) may consequently be combined and described by a function *independent of that variable across whose boundary they are adjacent*. For instance, going back to *Figure 10.7*, $\overline{A}\overline{B}C$ and $\overline{A}BC$ are adjacent across the B boundary, hence their sum is independent of the B variable and

$$\overline{A}\overline{B}C + \overline{A}BC = \overline{A}C$$

In other words, two cell (adjacent) groups are independent of *one* variable. The relationship could be easily found using logical algebra, of course, but the Karnaugh map makes the connection obvious by simple inspection.

We can now build on this. If the sum of products function contains another pair of terms, also adjacent across the B boundary and such that, after the above simplification, the resulting expression is seen to be adjacent to $\overline{A}C$ across either the A or B boundary, then all four terms (cells) may be combined into a single variable.

For example, ABC and $A\overline{B}C$ are adjacent across B and become AC. This in turn is adjacent to $\overline{A}C$ across A and hence

$$AC + \overline{A}C = C$$

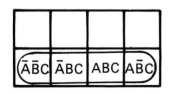

Figure 10.9

The map of *Figure 10.9* illustrates the situation as a loop drawn around the four cells concerned; in other words, four-cell groups are independent of *two* variables. What the map has told us, in fact, is that

$$\overline{A}\overline{B}C + \overline{A}BC + ABC + A\overline{B}C = C$$

Quite a simplification!

The process of combining the cells into two or four groups of adjacent cells in this manner is the basic operation on which the Karnaugh map simplifies or minimises the function.

(4) State which of the following pairs of cells are adjacent:

(a) $\overline{AB}C\overline{D}$, $AB\overline{CD}$ (b) $\overline{ABCD}$, $\overline{A}\,\overline{B}CD$ (c) $ABC\overline{D}$, $A\overline{B}\,\overline{C}D$

RULES FOR MINIMISATION

Before giving some typical examples of the use of the K-map, we can usefully summarise the rules which have been deduced in a general way over the preceding few pages.

Assuming that the function to be simplified is assembled into normal sum-of-product form, the steps to be taken are, in order:

1. Construct the K-map in the appropriate four, eight or sixteen cell form and enter the function in the form as described.

2. Examine the map for single 1s; adjacent *pairs* of 1s which cannot be included in a larger group; groupings of *four* 1s (which may include other previously looped as well as unlooped 1s) and group them after the manner shown in *Figure 10.9*. All 1s must be finally grouped in this way with no 'left-overs'.

3. Now write out the product terms for each loop formed.

4. Finally, OR the product terms to obtain the minimised expression in sum-of-product form.

This procedure may sound complicated, but with a little practice it is relatively easy to translate from the K-map into the required minimal function. Some worked examples now follow to illustrate first of all steps (1) and (2) of the above procedure.

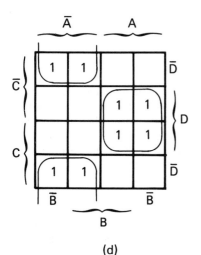

(a)

(b)

(c)

(d)

Figure 10.10

Example (5). Using appropriate K-maps, enter the following logical equations:

(a) $F = \overline{A}\overline{B} + A\overline{B} + \overline{A}B$ (b) $F = \overline{AB}\overline{C} + AB\overline{C} + \overline{A}BC + ABC$
(c) $F = AB + \overline{B}\overline{C} + A\overline{B}$
(d) $F = ABCD + \overline{A}\overline{C}\overline{D} + A\overline{B}D + \overline{A}CD + \overline{A}C\overline{D}$

In each case, group the cells as explained under rule (2) above.

We will work through each of these in turn.

(a) Since there are only two variables, A and B, we need a 4-cell K-map. This is drawn in *Figure 10.10(a)*. From the product terms of the given equation we can place 1s in the cells representing $\overline{A}\overline{B}$ (top left-hand cell). $A\overline{B}$ (bottom left-hand cell), and $\overline{A}B$ (top right-hand cell). The 1s thus marked can then be looped into two pair-groupings (with a common cell of $\overline{A}\overline{B}$) as illustrated.

(b) For this example we have three variables, so an 8-cell K-map is required. This is drawn in *Figure 10.10(b)*. This time the term $\overline{AB}\overline{C}$ goes into the second cell of the top row; $AB\overline{C}$ then goes immediately to its right; $\overline{A}BC$ goes into the second cell of the bottom row, with ABC immediately to its right. This completes the mapping. This time we can link the marked cells into a 4-cell cluster.

(c) Here again we need an 8-cell map to accommodate three variables. This is shown in *Figure 10.10(c)*. The three terms AB, $\overline{B}\overline{C}$ and $A\overline{B}$ seem at first to present a difficulty as all the cells have addresses involving three variables. There is no problem, however, if we notice that the *area* covered by the term AB is that area where A and B *overlap*, that is, it comprises *two* adjacent cells. Individually, these two cells are ABC and $AB\overline{C}$ which reduces

Karnaugh mapping 119

algebraically to AB. The term AB, then, is represented by the two cells of the third column. In the same way the term $\overline{B}\overline{C}$ is represented by the two outermost cells of the top row where $\overline{B}$ and $\overline{C}$ overlap; and the term $A\overline{B}$ occupies the two cells of the right-hand column. The completed map and the groupings are shown in *Figure 10.10(c)*.

(d) Four variables here so we need a 16-cell K-map. This is shown in *Figure 10.10(d)* but with a slight change in the layout. This alternative system of indicating the addresses has been put in at this point to illustrate a method which sometimes proves easier in filling in a larger map. The address of each cell is *exactly* identical with the layouts already used and illustrated originally in *Figures 10.1(c)* and *10.2(b)*. Check to make certain you understand that this is so. The map can now be filled in.

The first term ABCD is the easiest; it goes in the third cell along the third row. All the remaining terms have three variables each, hence they must occupy *pairs* of cells. $\overline{A}\overline{C}\overline{D}$ occupies the first two cells of the top row where $\overline{A}$, $\overline{C}$ and $\overline{D}$ overlap. $AC\overline{D}$ occupies the two right-hand cells of the second row. $AB\overline{D}$ occupies the two middle cells of the right-hand column. Finally, $\overline{A}C\overline{D}$ occupies the first two cells of the bottom row. Follow all this through for yourself on a rough piece of paper. The marked cells can now be linked into two groups, each of four cells.

Try mapping the following examples on your own, the first of these directly from the truth tables.

(6) Draw K-maps for (a) a two-input AND gate, (b) a two-input OR gate, (c) a two input NAND gate.
(7) Map the functions and group as necessary:
    (a) $\overline{A}\overline{B}\overline{C} + \overline{A}B\overline{C} + \overline{A}\overline{B}C + \overline{A}BC$
    (b) $\overline{A}\overline{B}\overline{C} + \overline{A}BC + AB\overline{C} + A\overline{B}C$
    (c) $\overline{A}\overline{B}C + \overline{A}B\overline{C} + A\overline{B}\overline{C} + ABC$
    (d) $\overline{A}BC + \overline{A}B\overline{C} + AB\overline{C} + A\overline{B}C$
    (e) $\overline{A}\overline{B}\overline{C}\overline{D} + \overline{A}BC\overline{D} + AB\overline{C}\overline{D} + A\overline{B}\overline{C}\overline{D} + ABC$

READING THE K-MAP

We have now had some practice in filling the K-map and grouping the marked cells. We want now to read what the K-map tells us about minimisation. So we turn to steps (3) and (4) from the above set of procedures, and again illustrate by way of examples.

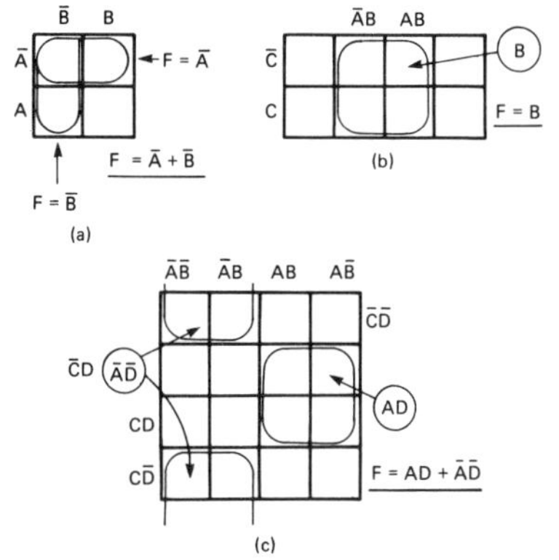

Example (7). Using the maps obtained from Example (5) above, minimise the expressions concerned by inspection of the appropriate map.

For example (a) the given equation was $AB + A\overline{B} + \overline{A}B$ and the mapping was drawn in *Figure 10.10(a)*. For convenience of explanation, the map is redrawn in *Figure 10.11(a)*. If we look at the looped section in the upper row, we see that $\overline{A}$ is independent of B; the loop crosses the B-boundary along the $\overline{A}$ row. Hence this loop represents just the term $\overline{A}$. Similarly, for the left-hand loop, we notice that $\overline{B}$ is independent of A; the loop crosses the A-boundary along the $\overline{B}$ column. Hence in this case the loop represents just the term $\overline{B}$. To obtain the minimal function, we

Figure 10.11

OR these terms and so obtain $F = \bar{A} + \bar{B}$. This is the minimal form of the original equation. You should recognise it as the expression for a two-input NAND gate, since $\bar{A} + \bar{B} = \overline{AB}$ by De Morgan's rule.

In example (b) the given equation was $\overline{AB}\bar{C} + AB\bar{C} + \bar{A}BC$ + ABC and the mapping was given in *Figure 10.10(b)*. The map is repeated in *Figure 10.11(b)*; the 4-cell grouping must now be interpreted. Notice that the loop crosses both the A and C boundaries, the expression we want is therefore independent of both A and C. The 4-cell cluster is thus just equal to B. This is the minimal state of the given equation, or $\overline{AB}\bar{C} + AB\bar{C} + \bar{A}BC + ABC = B$.

We skip now to example (d) where *Figure 10.10(d)* showed us the map of $ABCD + A\bar{C}\bar{D} + A\bar{B}D + \bar{A}CD + \bar{A}C\bar{D}$. Looking at the equivalent mapping in *Figure 10.11(c)* we see that the 4-cell cluster on the right centre crosses the B and C boundaries and is contained within the A and D address zones. Hence it represents just the term AD. For the other (split) 4-cell cluster, this crosses the B boundary and the (edge) C boundary and is contained within the $\bar{A}, \bar{D}$ address zones. Hence it represents just the term $\bar{A}\bar{D}$. The minimal function therefore is $AD + \bar{A}\bar{D}$ or $ABCD + A\bar{C}\bar{D} + A\bar{B}D + \bar{A}CD + \bar{A}C\bar{D} = AD + \bar{A}\bar{D}$.

Other forms of the equation are possible, but none will contain fewer or simpler terms.

You will no doubt have noticed that we have not looked at the map shown in *Figure 10.10(c)*. Try this one on your own.

(8) Show that the minimal expression for the mapping of *Figure 10.10(c)* is given by $A + \bar{B}\bar{C}$.

APPLICATION TO CIRUIT DESIGN

Most problems associated with logic circuitry stem from the need to realise a desired function from a combination of gates. We have mentioned earlier how it is possible to make up such a combination and then simplify it into an equivalent system using both a smaller number and a smaller selection of gates. The approach always is to proceed from a consideration of the required function to a truth table and then to a logical expression of the function. This expression can then be minimised by algebra or by a Karnaugh mapping. Finally, the circuit realisation of the minimised function is assembled and tested. We conclude this unit section with an example of the above procedure.

Example (9). The output of a logical system is required to be given by $F = \overline{AB}C + A\bar{B}C + AB\bar{C} + ABC$. Using *any* logic gates, design a circuit which would provide this output. By using a Karnaugh mapping, simplify the system so that a small number of *identical* gates may be used.

We start off with no restrictions on the number or type of gates we may use. From the logical equation, we have to invert one of the variables in each of three of the terms, then AND the four combinations of the terms, and finally OR all four terms. This can be done with three NOT gates, four AND gates and an OR gate; the circuit based on this reasoning is shown in *Figure 10.12*. The intermediate states are marked on the diagram and you should

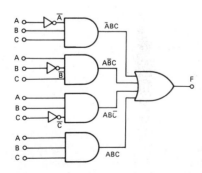

Figure 10.12

Karnaugh mapping

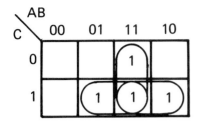

Figure 10.13

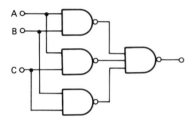

Figure 10.14

have no difficulty in tracing the signals through the system. If this circuit was built using the 7400 series of TTL gates, you would have to use more packages than the eight which the diagram suggests. The NOT gates are no problem, three of these can be found in the 7404 package (which contains six altogether). The 7411 is a triple 3-input AND package, but as we require four 3-input AND gates, you would need two 7411s. The OR gate is a 4-input job; it is left as an exercise to find suitable gates and *arrangements* from the 7400 series to accomplish this part of the logic.

Let us see if we can simplify the system. The output expression from *Figure 10.12* is plotted on a Karnaugh map in *Figure 10.13*, and three looped pairs are obtained. From an inspection of these, we find that the minimised function is $F = AB + AC + BC$. This is going to be easier to put into circuit form. It could obviously be achieved with three AND gates followed by an OR gate. If you go to the 4000 series of gates, a 4081 and a 4075 would do the trick. However, if we are to use identical gates, then the output we want suggests a double inversion of each product pair of terms, and this can be done with four NAND gates as shown in *Figure 10.14*. A couple of 7400 packages would do the job, or a 7400 with a 7410 could be used.

PROBLEMS FOR SECTION 10

(10) Map the following logical expressions and find the minimal sum of products form for each of them:
(a) $ABC + AB\bar{C}$
(b) $\bar{A}B + AC + BC$
(c) $\bar{A}BC + AB\bar{C} + ABC$
(d) $\bar{A}BC + A\bar{C} + ABC$
(e) $\bar{A}\bar{B}C + \bar{A}BC + \bar{A}B\bar{C} + AB\bar{C}$
(f) $AC + B\bar{C} + \bar{A}BC + ABC$
(g) $A\bar{B}\bar{C} + \bar{A}BC\bar{D} + \bar{A}CD + \bar{B}C\bar{D} + AD$

(11) A Karnaugh map for a four-variable function has sixteen cells. The four corner cells of this map are mutually adjacent. Evaluate the combination of the four corner cells. (Treat them as a 4-cell cluster.)

(12) From the Karnaugh map shown in *Figure 10.15*, write down the logical terms represented by the groupings P, Q, R and S.

(13) What are the advantages of using only one type of gate in a logic circuit system? Which are those gates from which all other logical gates can be derived? Show that the circuit shown in *Figure 10.16* can be replaced by using four NAND gates.

(14) Using the truth table given below, write down the logical equation at the output in sum of products form. Simplify this equation by use of a Karnaugh map and design a circuit system which would provide this output. (You may use AND, OR and NOT gates.)

A	B	C	D	F
0	0	0	0	1
0	0	0	1	1
0	0	1	0	1
0	0	1	1	1
1	1	1	0	1
1	1	1	1	1

All other combinations give F = 0.

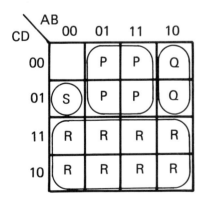

Figure 10.15

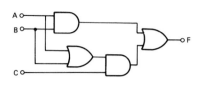

Figure 10.16

(15) You have a quad two-input NAND package (a 7400). Show how you could use this package to make a two-input EX-OR gate, using all four NAND gates.

(16) Three civilians (A, B, C) and an army officer (D) are seated in a compartment of a train. The window will be open if the officer and at *least* two of the civilians vote to have it open.

(a) Write out a truth table for the window OPEN situation, taking a YES vote to equal 1 in the table and OPEN equal to 1.

(b) Use a Karnaugh map to produce the minimised expression for the situation.

(17) Use a K-map to simplify $F = B\overline{C} + \overline{A}BC + A\overline{B}\overline{C} + AC$. Show how the minimised expression can be implemented using two NOR gates only.

11 Sequential logic systems

Aims: At the end of this Unit section you should be able to:
Distinguish between combinational and sequential logic systems.
Understand and write truth tables for flip-flops and latches.
Analyse the operation of integrated R-S, D-type and J-K flip-flops.
Construct a shift register using J-K flip-flops.
Construct a binary counter using J-K flip-flops.

We come now to what are known as sequential logic systems in which the circuits, of which registers and counters are examples, work with timed sequences of clock (or enabling) pulses. This distinguishes these systems from the purely combinational gate arrangements we have discussed in earlier chapters. There the outputs of the gates were determined by the existing inputs; in sequential systems the outputs may be affected by past inputs as well as those actually present at that time. The basic circuit module from which most sequential logic is derived is the bistable multivibrator or *flip-flop*. In this Unit section we will consider the operation of the flip-flop first in its discrete form and then in its integrated form, and some of the many applications it has in the design of digital systems.

A TRANSISTOR FLIP-FLOP

Digital systems are almost always built around registers which are groups of flip-flops. A flip-flop is a two-state element which can be set in either of the two states and remain in that state until some other input signal changes the state.

The operation of a flip-flop depends upon the fact that a transistor is both a switch and an amplifier. *Figure 11.1* shows a single transistor set up as an elementary flip-flop. The base bias derived from the potential divider R_1 and R_2 is such that the transistor is normally cut off; the collector potential is then high. If a positive potential is applied to input A, the base-emitter junction will be forward biased and the output will go low. This is simply a straightforward transistor switch or inverter.

However, there is an alternative input terminal we might make use of; if we make the potential at the positive end of R_1 *more* positive, the transistor could be switched on without anything being applied to input A. Hence we could get the transistor to switch if either input A *or* input B were made positive. In conjunction with the normal inversion (NOT) function of the transistor, the system is essentially a NOR gate. By using two such gates, a flip-flop can be created. The circuit for this is shown in *Figure 11.2*.

Here we have two transistors, each individually set up in the form of *Figure 11.1*; the resistors have been numbered to make this clear. The output of each transistor feeds back to the input of the other but in such a way that each output is effectively the input B point of *Figure*

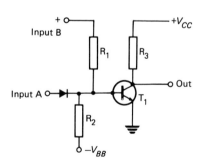

Figure 11.1

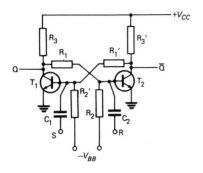

Figure 11.2

11.1, and the input points now marked R and S correspond to the input A point of *Figure 11.1*.

Suppose that both transistors are conducting and are drawing equal collector currents. Because the circuit appears symmetrical in its layout such a condition may seem possible, but actually it is impossible. No two transistors have precisely identical characteristics and no resistors have precisely the same ohmic values. Even if we select the components from close tolerance values, there are going to be electrical differences between one side of the circuit and the other. Hence, one collector current is going to be greater than the other. So assume that the collector current of T_1 is slightly greater than that of T_2 and increasing. The voltage drop across R_3 will increase and the collector potential will fall. This fall will be coupled through resistor R_1 to the base of T_2 and T_2 will experience a drop in its collector current and a consequent rise in its collector potential. This rise is in turn coupled back through resistor R_1' to the base of T_1 where it assists the increase in collector current already taking place. The effect is cumulative and the action continues until transistor T_1 is driven into saturation (fully conducting) and T_2 is cut off (non-conducting). Nothing further can then happen. Although this description has taken a minute or two to describe, the action takes place very rapidly after switch-on, usually in a fraction of a microsecond.

The circuit now rests in a *stable* state and will remain in this state indefinitely. Transistor T_1 is saturated and its collector is at (near) zero volts. This condition, in conjunction with the negative bias applied to its base via R_2, ensures that T_2 is firmly cut off. With T_2 cut off, its collector potential is large and positive and this is sufficient to overcome the effect of the negative bias via R_2' on the base of T_1. Hence T_1 remains conducting.

If we had started this analysis by assuming that the collector current of T_2 was greater than that of T_1, and increasing, the regenerative action would have swung in the opposite direction, with transistor T_2 being fully switched on and T_1 cut off in the stable state. The action would otherwise have been identical. There are therefore two stable states possible with this circuit: either T_1 is ON and T_2 is OFF, or conversely. So the outputs at the collectors are complementary; if one of them is Q, the other is $\bar{Q}$.

Because it has two stable states, this circuit is known as the bistable multivibrator or, in digital applications, the R-S (RESET-SET) flip-flop.

An Experiment

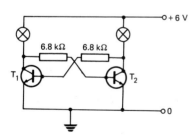

Figure 11.3

A simple experiment will demonstrate the operation of this basic flip-flop. Build the circuit shown in *Figure 11.3* using a pair of BC107 or BC108 transistors. The bulbs, which act as collector loads and which indicate which of the transistors is ON and which is OFF, should be low current types, typically 6 V at 60 mA. When you switch on, one or other of the bulbs will light. If you switch on and off a number of times you will most likely find that the same bulb lights up each time; this is because the circuit 'unbalance' favours that particular side.

If now you momentarily short out to the negative (earth) line the base of the transistor which is ON, the circuit will switch states and remain in this new condition until you repeat the shorting procedure. What happens is that when you short out the base of the ON transistor, its collector current falls to zero and its collector potential rises to 6 V. The OFF transistor now receives a base current flowing through its base resistor and switches ON. Its collector consequently falls to near zero

Sequential logic systems 125

volts, thus preventing the other transistor (now OFF) from receiving any base current even when the short circuit is removed. Make a note of the fact that the circuit will not switch states if the already OFF transistor is shorted.

TRIGGERING THE FLIP-FLOP

Go back to the circuit of *Figure 11.2* and assume that T_1 is switched ON and T_2 OFF, a stable state that gives the output at the collector of T_1 as low, or logical 0. Now suppose a negative pulse to be applied to the base of T_1 by way of the S (or SET) terminal. This negative pulse will switch T_1 OFF and the circuit will change state in the manner already described. In this condition the circuit is again perfectly stable and will remain so even when the *trigger* pulse is removed. The output at the collector of T_1 is now high, or logical 1. Any further negative pulses applied to the SET terminal will have no effect since T_1 is already switched off. However, a negative pulse applied to the R (or RESET) terminal will switch T_2 OFF and a change of state will immediately follow. The conventional way of representing this elementary bistable flip-flop is shown in *Figure 11.4* where the outputs are designated Q and $\overline{Q}$.

Triggering the R–S flip-flop in this way would not be very convenient in practice because it would be necessary to interchange the S and R inputs for each required change of state. This difficulty can be overcome by modifying the circuit to that shown in *Figure 11.5*. Here the setting and resetting operation is carried out automatically, each negative pulse applied to the single input terminal causing a change-over. Two diodes have been added to the basic circuit and these direct or steer the input pulses to the appropriate transistor base. For this reason they are known as *steering* diodes. Assuming that T_1 is switched ON, output Q is consequently at logic level 0 and output $\overline{Q}$ is at logic level 1.

Now looking at diodes D_1 and D_2, under our assumed conditions D_2 is reverse biased and D_1 is forward biased. The anode of D_2 connects

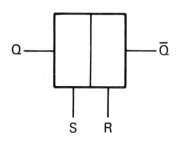

Figure 11.4

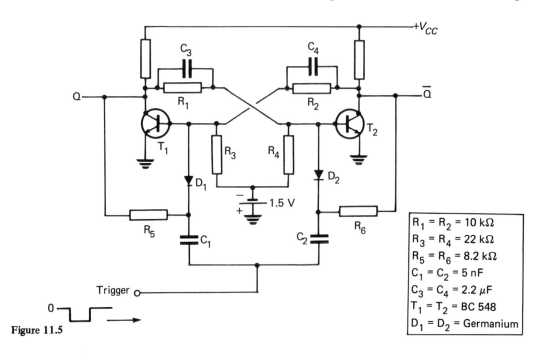

Figure 11.5

via resistor R_1 to the collector of T_1 which is low (near 0 V) but its cathode connects via resistor R_6 to the collector of T_2 which is high (V_{CC}), hence D_2 is reverse biased. The anode of D_1, however, connects via resistor R_2 to the high collector of T_2 while its cathode is returned via resistor R_5 to the low collector of T_1, hence D_1 has a small forward bias. When a negative trigger pulse is fed to the diode cathodes through capacitors C_1 and C_2, the conducting diode D_1 passes the negative change on to the base of T_1 but diode D_2 remains non-conducting. As T_1 switches OFF its collector potential rises, T_2 switches ON and its collector potential falls, so reducing the base current of T_1. The circuit rapidly changes state and the output logic levels reverse, Q now becoming $\bar{Q}$, and $\bar{Q}$ becoming Q. The bias conditions on the steering diodes are now reversed also, so that the following negative trigger pulse cuts off T_2 and the circuit reverts to its original state. The capacitors C_3 and C_4 shown in the diagram are normally included to assist in the rapidity of switch-over between the two states.

Figure 11.6 shows the waveforms of the input and output signals for a succession of trigger pulses. You should particularly notice that the bistable changes state on the negative-going edge of the trigger pulses. If the output is taken from one collector only, one output pulse is obtained for every *two* input pulses. Thus the circuit functions as a binary divider. In this way, any input frequency may be divided by any power of two (four, eight, sixteen, etc.) by combining the requisite number of bistables in cascade.

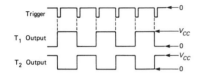

Figure 11.6

(1) In the experimental circuit of *Figure 11.3* can you think of an alternative method of getting the bistable to switch state, again by shorting two points together?

(2) Using the values given in the list alongside *Figure 11.5*, make up the circuit, replacing the collector resistors by low-current bulbs (6 V, 60 mA), as indicators. Feed in trigger pulses from a low-frequency square-wave oscillator, negative-going, with a frequency of a few hertz. Note that the circuit output pulses are half the frequency of the input pulses. (A V_{CC} supply of 9 V is suitable).

INTEGRATED SYSTEMS

The basic flip-flop so far described is essentially nothing more than a pair of cross-coupled NOT gates: *Figure 11.7(a)*. If the gates are replaced by a pair of two-input NAND gates or a pair of two-input NOR gates, a controllable bistable becomes available in integrated circuit form: see *Figure 11.7(b)* and *(c)* respectively.

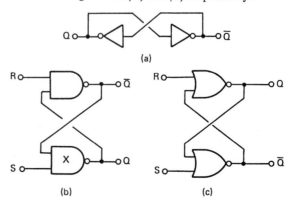

Figure 11.7

Strictly these two circuits are what are known as *latches*, which will be explained as we go along, but they behave in much the same way as the R–S bistable already discussed and are commonly referred to as R–S bistables.

The SET and RESET inputs connect to one input on each gate, the other inputs being connected to the opposite outputs in cross-coupling. We will analyse the two-NAND system, and after that you should be able to work out the operation of the two-NOR circuit for yourself. Referring to *Figure 11.7(b)*, suppose R and S to both be held at logic 1, then the NAND gates simply act as inverters and the system holds its output state indefinitely, i.e. Q and $\bar{Q}$ are 'stored' in whatever condition they were in already and the circuit is said to be in its latched state. If S now falls to logic 0 while R remains at logic 1, output Q will switch to logic 1 because NAND gate X has an input at logic 0. In this condition the circuit is said to be SET. Returning S and R both to logic 1 will not reset the bistable and it will remain latched with Q = 1, $\bar{Q}$ = 0. Similarly, if R drops momentarily to logic 0 while S remains at logic 1, output Q will switch to logic 0, and hold there when R is returned to logic 1. The truth table for this R–S function is shown in *Table 11.1*. By convention, Q = 0 and $\bar{Q}$ = 1 in the RESET state.

Table 11.1

	S	R	Q	$\bar{Q}$
Latch →	1	1	Q	$\bar{Q}$
Set →	0	1	1	0
Reset →	1	0	0	1
Indeterminate	0	0	•	•

You will notice that the output state when both S and R are at logic 0 is described as indeterminate. This follows from the fact that *any* low input on a NAND gate makes the output high, hence for S = R = 0 both Q and $\bar{Q}$ would be high. But this is a contradiction of the definition that Q and $\bar{Q}$ are complementary; in other words, the output cannot be determined if R and S are both low.

The waveforms shown in *Figure 11.8* illustrate how the circuit operates. Suppose both inputs to be initially high and Q low, then if S goes momentarily low, Q will go high. The pulse width, provided it exceeds a certain minimum, is not important. The high on Q is now latched and held there and the circuit is SET. The only way to RESET the circuit (get Q back to low) is to put the R input momentarily low after the S input has gone back high.

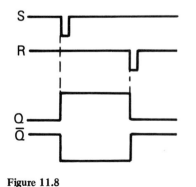

Figure 11.8

(3) Draw up a truth table for the circuit of *Figure 11.5* covering a complete transition of the output cycle.

(4) Analyse the bistable latch illustrated in *Figure 11.7(c)* and draw its truth table. Hint: use the analysis above for the NAND circuit as a guide.

THE CLOCKED R–S FLIP-FLOP

The addition of extra circuitry to the basic R–S flip-flop will not affect its fundamental operation but will give it added flexibility. The circuit of *Figure 11.7(b)* may be modified to bring it into line with the triggered bistable of *Figure 11.5* by adding a pair of OR gates and a NOT gate.

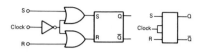

Figure 11.9

By this means, the facility of separate R and S inputs is retained while providing a common trigger input. In this regard, trigger pulses, known as *clock* or *enabling* pulses, are generated by (often) a high-accuracy pulse generator. This clock input allows the input conditions to be transferred to the output on one edge of the clock. The circuit is shown in *Figure 11.9*, together with its conventional symbol.

This circuit can only change state when the clock input is high. If the clock input is low, the output of the NOT gate is high, hence each OR gate has an input at logic 1 and their outputs must also be 1. Both the S and R inputs to the basic flip-flop are therefore at 1 and, from the truth table, the Q and $\bar{Q}$ outputs are held. When the clock input goes high, however, the output from the NOT gate goes low and the OR gates each have an input at logic 0. The S and R on the flip-flop then depend only on the logic levels at the *external* S and R inputs; the state in which the flip-flop will set itself will then be in accordance with the truth table. The clocked R–S flip-flop therefore responds to the R and S inputs only when the clock input is high. For this reason, this logic level is referred to as the enabling pulse.

THE D-LATCH

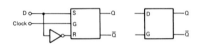

Figure 11.10

The D (or data) flip-flop overcomes the problem of the basic R–S model having an indeterminate state when both the S and R inputs are at logic 0. The difficulty can be overcome by putting an inverter between the R and S inputs as shown in *Figure 11.10*. This circuit is available in quad format in the 7475 TTL or the 4042 CMOS. As the truth table of *Table 10.2* shows (try analysing the circuit for yourself), the Q output follows the D input as long as the clock logic is high. The logic state present on the D input just before the clock goes low will be latched on the output.

Table 11.2

D	Clock	Q	$\bar{Q}$
.	0	Q	$\bar{Q}$
0	1	0	1
1	1	1	0

One of the main uses for the D-latch is to hold or *remember* a certain binary number or a displayed output on a counter unit while the actual count sequence continues. For normal counting, the clock input is held high (connected to $+V_{CC}$); for the latched output it is connected to low (earth line). The symbol for the D-latch is shown alongside *Figure 11.10*. The clock input is usually designated G.

The D flip-flop is sometimes confused with the D latch. Whereas in the D latch the output Q follows the D input as long as the clock is high, in the flip-flop the Q output consists only of data which is present on the D input at the time of a *positive transition* on the clock input. This distinction is not particularly important to us at this stage because the terms 'flip-flop' and 'latch' are both used in a general and perhaps haphazard sense, but it is mentioned to show that such a distinction exists.

THE J–K FLIP-FLOP

This very versatile kind of flip-flop can be looked on as a combination of R–S and D-type flip-flops with additional control, these being *preset* and *clear* input points. These additions enable the flip-flop to be set up in a known state, say on switch-on, so that the initial conditions are established prior to full operation. When not required, these inputs are held low and do not influence the normal operation of the circuit.

A circuit arrangement (one of several possible) is shown in *Figure 11.11*. Two flip-flops are involved, one being the *master* and the other the *slave*. The master responds to data from the J and K inputs while the clock pulse is high (as for the clocked R-S flip-flop already described), but this is only transferred to the output (slave) flip-flop as the clock goes low. The slave is a D-type latch. The clock input feeds directly to the master but by way of an inverter to the slave. The slave consequently receives a high clock input when the true clock is low, and the Q output is waiting to respond to any change in the D input. There is no possibility of any such change, however, all the time the clock is low because the master flip-flop is in a stable state reached when the clock was last high.

When the clock does go high, the Q output of the master responds to the R and S inputs but the D input on the slave is unresponsive because of the clock input now being low via the inverter. When the clock returns to low, the Q output of the master is held in the new state and, simultaneously, the slave clock input goes high and enables it to respond to this new state. Hence the J-K flip-flop changes its output state on the falling edge of the clock pulse, in that the master receives the information on the positive edge and transfers it to the slave on the falling edge. Since the clock does the triggering, the inputs themselves need not be in pulse form. To summarise: the operating sequence is that first the slave is isolated from the master, then the J and K inputs are entered into the master. The J and K inputs are then disabled and finally the information is transferred from the master to the slave.

The truth table for the J-K flip-flop is shown in *Table 11.3* where Q_o indicates the present state of the output. With low inputs at J and K, the clock pulse has no effect and the existing state is maintained. With input J high and K low, any clock pulse sets the output high; with J low and K high, any clock pulse resets the output low. Notice that when *both* inputs go high, the output changes state. This is the divide-by-two action produced by the discrete circuit of *Figure 11.5* and is known as 'toggling'.

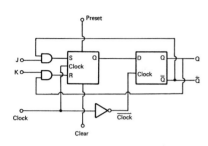

Figure 11.11

Table 11.3

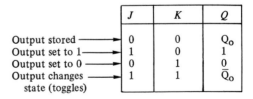

	J	K	Q
Output stored	0	0	Q_o
Output set to 1	1	0	1
Output set to 0	0	1	0
Output changes state (toggles)	1	1	$\bar{Q}_o$

J-K flip-flops (as are R-S and D flip-flops) are available in integrated circuit packages, the 7474 being a dual D-type latch and the 7476 being a dual J-K flip-flop with preset and clear facilities. *Figure 11.12(a)* and *(b)* show the internal arrangements of these two packages. The preset input allows either a 1 or a 0 to be stored in the system; the clear input allows all old data to be cleared to 0, an example of which you will recognise in the 'Clear' button of a pocket calculator.

130 *Sequential logic systems*

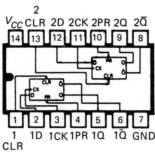
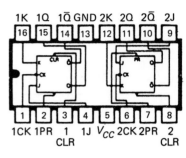

(a) 7474 dual D—type flip-flop

(b) 7476 dual J-K flip-flop

Figure 11.12

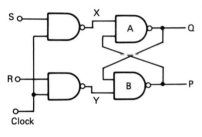

Figure 11.13

Example (5). Figure 11.13 shows a possible arrangement of a flip-flop that will store a single bit of information. Use logical algebra to show that the S input will be stored when a clock pulse is applied.

Let the input, output and intermediate levels be marked as shown in the diagram. Then when the clock pulse is not present the outputs X and Y must be high, since one input to each NAND gate is low. The inputs to the flip-flop itself are therefore Q and 1 on gate B, and P and 1 on gate A. Therefore

$$P = \overline{Q} \quad \text{and} \quad Q = \overline{P}$$

This is a stable condition and the output will remain at Q indefinitely.

When the clock pulse is applied, $C = 1$, $X = \overline{S}$ and $Y = \overline{R}$;

$$\therefore Q = \overline{\overline{S}.P}$$
$$P = \overline{\overline{S}.Q}$$
$$\therefore Q = \overline{\overline{S}.(\overline{S}.Q)} = S + S.Q$$
$$= S(1 + Q) = S$$

Hence the output is S when the clock pulse is applied.

This example introduces us to the next section.

SHIFT REGISTERS AND COUNTERS

Shift registers are employed to store information for a great number of purposes: arithmetic calculations and operations, program control, buffers in the movement of data, to name only three. Essentially, since a register is a memory unit, it is based on the flip-flop, and a series of J-K flip-flops (using the 7476 TTL package, for example) can be used to store a series of binary digits. Digital signals are fed into the first stage of the series and are moved in sequence along the chain of flip-flops until a complete number (or word) is stored. Registers are used extensively in calculators and computers and in data processing for parallel-to-series (or conversely) conversion of information.

Figure 11.14 shows the connection of four J-K flip-flops to make a 4-stage shift register: a register, that is, which will store a 4-bit binary

Sequential logic systems 131

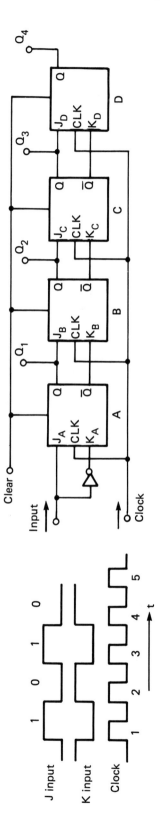

Figure 11.14

number, anything from 0000 to 1111 or 0 to 15 in denary notation.

To illustrate the action we will enter the binary number 1010 (=10 denary) into the register. First of all, all the flip-flops are reset to 0 by momentarily connecting the CLEAR terminal to logic 1. The number 1010 is represented by the logic levels 1, 0, 1 and 0 and is fed one bit at a time into the J terminal of flip-flop A. The K terminal receives the complement of the J input by way of the inverter. At the first clock pulse, $J_A = 1$ and $K_A = 0$, so that Q_1 is set and the first digit has been stored in the A flip-flop. At the second clock pulse, J_B is found to be 1 so B is reset and this logic level is transferred to output Q_2. At the same instant, flip-flop A now has a logic 0 on J_A (and 1 on K_A) so that Q_1 switches to 0 on this second clock pulse. Two of the digits are now stored in the register. After four such clock pulses, the full number 1010 has been entered into the register. If LED indicators are connected to the Q output points, they will display the stored number by lighting where the output logic is high.

A shift register of this kind not only stores a piece of information but converts the input *serial* application of the bits into *parallel* output mode. Aside from its storage properties, this facility makes the register a data-conversion device. The input sequence is clearly fed one digit after another; the output is available at Q_1, Q_2, Q_3 and Q_4 as simultaneously available digits. Of course the data can only be read out in this way when the register is not carrying out its internal shifting of the bits. Computers usually process their information in parallel form, but their inputs are in serial form, coming as a time sequence of 1s and 0s from the keyboard, electric typewriter, magnetic tape or other source. *Figure 11.15* shows how two shift registers might be employed to connect a keyboard to a data terminal over a single wire link (such as a telephone line). It would not be possible to transmit the bits in parallel form over such a link because each bit would require its own separate line. Hence it is necessary to convert any input to the link into a serial mode of transmission and reconvert it back to parallel mode at the output of the link. The system is kept in step (or synchronised) by using a common clock signal. As drawn in the figure, the system allows 4-bit words to be transmitted from the sending register to the receiving register in serial form. The input from the keyboard and the output to the terminal, however, require to be in parallel form, so the registers are being used as parallel-in/serial-out (PISO) at the sending end and as serial-in/parallel-out (SIPO) at the receiving end of the link. The 74164 is a serial/parallel data converter.

The parallel facility is obtained by using the PRESET input on each flip-flop (not shown in *Figure 11.14*). If, once the four digits are stored in the register, additional digits are introduced, those already in store are pushed out at the far end of the chain and lost. If data is entered at the preset inputs in parallel form, the clock train will feed out the digits in serial form, so making the register a PISO converter. The four bits, then, of *Figure 11.15* are first loaded from the keyboard into the sending register in parallel; then, on receipt of the clock signal, both sending and receiving registers shift four places. The information is pushed out of the sender and enters the receiver where it is read out in parallel by the same clock signal and transferred to the terminal.*

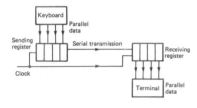

Figure 11.15

*The systems and protocols associated with serial transmissions over the telephone network are covered in a companion book, *Electrical and Electronic Principles 2.*

Counters We turn now to a sequential logic unit which derives immediately from the registers we have just discussed. This is the counter.

Counting is essentially a matter of being able to add up. If we can add one digit at a time, we are counting. By using the binary system of numbering, counting by electronic means is simplified: one binary counter need store only two digits, 0 or 1, in any column, rather than 10 digits, from 0 to 9, as would be necessary in a decimal (denary) counter. Counters, irrespective of what they might appear to be, basically can all be looked on as serial registers with a single input and a parallel output from each of the separate flip-flops making up the register. The function of any counter is to count the pulses as they arrive at the input. There are obviously many applications for such counting systems: batch counting on a production line, the measurement of time intervals and the measurement of frequency, again to name only three.

The basic element in most binary counters is the familiar J-K flip-flop, though a D-type will also serve. A J-K flip-flop with the J and K inputs tied high, or a D-type with the $\bar{Q}$ output connected to the D input will toggle: the output state will change for each input pulse. Since two input pulses produce one output pulse, the element is a divide-by-two unit as we have already seen. *Figure 11.16* shows how four J-K flip-flops might be assembled in cascade to form a 4-bit binary

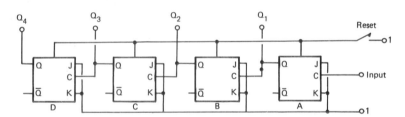

Figure 11.16

counter. Counters of this sort are called *asynchronous*, this term merely implying that the counting sequence is not synchronised to or by anything else, or compared with an external frequency to check its accuracy. Errors can occur, particularly when the counting rate is high because of the time delay experienced by the signal as it passes through the flip-flops, but this aspect need not concern us at this stage.

In the circuit, the J and K inputs are both taken high so that each flip-flop will toggle. As for the register, the outputs can be set to zero by momentarily connecting the CLEAR terminals high, after which the system is ready to count. Notice that the pulse inputs which are fed in at the CLOCK terminal of the first flip-flop, have been arranged in the diagram to be on the right-hand side. This has been done to get the output information in the 'right order'. This is also applicable to shift registers. The count is available as a four-digit binary number on the (parallel) Q_1, Q_2, Q_3 and Q_4 outputs, and of these Q_1 is the least significant bit (LSB).* As the diagram is drawn, therefore, the digits are

---

*The least significant bit is the number 'on the right'. In the denary number 3675, for example, 5 is the LSB. It represents the 'units'. The 7 represents 70, the 6 represents 600 and the 3 represents 3000. This left-hand number is the most significant digit.

134 *Sequential logic systems*

Table 11.4

Input pulse	Q_4	Q_3	Q_2	Q_1
0	0	0	0	0
1	0	0	0	1
2	0	0	1	0
3	0	0	1	1
4	0	1	0	0
5	0	1	0	1
6	0	1	1	0
7	0	1	1	1
8	1	0	0	0
9	1	0	0	1
10	1	0	1	0
11	1	0	1	1
12	1	1	0	0
13	1	1	0	1
14	1	1	1	0
15	1	1	1	1

in their correct order of significance. The binary outputs for the input pulse numbers from 0 to 15 are shown in *Table 11.4*. From this we see that, because of the divide-by-two function of each flip-flop, each time any output goes from 0 to 1 the following flip-flop toggles to the opposite state. It consequently takes two 1-to-0 transitions at the input to cause output Q_1 to change once; four transitions to cause output Q_2 to change once; eight transitions to cause output Q_3 to change once, and sixteen transitions to cause output Q_4 to change once. Thus after each input pulse the circuit gives a binary representation of the number of pulses received up to that point; so the circuit is counting in binary code. After the sixteenth pulse (1111) the stages reset to zero and the count begins again. This is a scale-of-sixteen counter. If numbers greater than 1111 (decimal 15) have to be counted, more stages must be added. As we have already noted, n stages will give a scale of 2^n counter.

(6) It is desired to count up to a maximum of decimal 512 using a scale-of-sixteen binary counter. How many flip-flop stages are necessary?

Now counting in blocks of powers of two (or modulo-2) is not a particularly convenient way from the point of view of reading out the count. Of course, it is perfectly possible to be able to calculate in binary and many people can do this with the same fluency as the rest of us count in the decimal or modulo-10 system. If the counter can be modified to count only up to ten before resetting to zero, the counting pattern will at least be a little more familiar, although the output total will still be seen in binary form. A modification of the basic counter is shown in *Figure 11.17*. This is only one of several alternatives but it is probably the easiest to follow. For clarity, the J and K inputs are omitted. The binary equivalent for decimal ten is 1010; this has four digits and can be handled by a four-input logic gate. In the figure an AND gate has been selected. Notice that the four inputs to this gate are

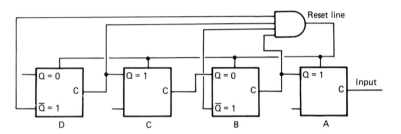

Figure 11.17

taken from the Q outputs on the A and C flip-flops and from the $\bar{Q}$ outputs on the B and D flip-flops. When the count reaches 1010 at the four Q terminals respectively, the inputs to the AND gate will all be simultaneously at logic 1. The output of the AND will then change to 1. This output applied to all four reset (CLEAR) inputs on the flip-flops will cause the counter to reset to 0000 on the count of ten, after which the cycle will recommence. We now have a decade or modulo-10 counter.

The system can be easily adapted to other counting cycles by changing the inputs to the AND gate, noting that A is the most significant digit. For example, a scale-of-twelve counter can be made by noting that the binary equivalent for decimal twelve is 1100. The inputs to the AND gate are therefore taken from the Q outputs on the A and B flip-flops and from the $\bar{Q}$ outputs on the C and D flip-flops.

(7) How would you adapt the circuit of *Figure 11.16* to count in (a) a scale of six, (b) a scale of eleven?

Integrated Counters   Binary counters are readily available in integrated circuit form and these usually include additional functions so that different counting sequences are available to the user. They are also available to drive seven-segment read-out display devices directly or by way of a suitable decoder so that the output is presented in decimal form. Any pocket calculator is an immediate example.

We will examine one or two of the more popular TTL and CMOS packages available for counting purposes.

136 Sequential logic systems

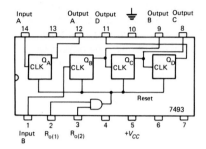

Figure 11.18

The 7493 TTL integrated circuit is a self-contained 4-bit binary counter and is similar to the circuit shown in *Figure 11.16* with an additional 2-input AND gate included to make provision for counting sequences other than scale of sixteen. *Figure 11.18* shows the connections and the circuit arrangement of the 7493. All four flip-flops have a common reset line fed from the output of the AND gate. Also flip-flop A has its output brought separately to pin 12 and not tied to the input of flip-flop B as the other three flip-flops are connected. Flip-flop A can therefore be used separately as a straightforward divide-by-two (using pins 14 and 12) while input B (pin 1) and outputs B, C and D will provide a two-, four- or eight-times division, respectively. For use as a 4-bit counter, output A is connected to input B. Notice here that output A is the least significant digit.

Resetting the outputs back to zero takes place through the included AND gate whose input is brought out on pins 3 and 4 and indicated as $R_{o(1)}$ and $R_{o(2)}$ respectively. This is an arrangement similar in form to that described under *Figure 11.17* earlier. To reset the counter, the reset line must be taken momentarily high; this will occur when both inputs to the AND gate are high. Counting can only take place when at least one of the AND inputs is low.

To count to a scale of ten (binary 1010), the inputs to the AND gate are taken from those outputs which are high on the count of ten, that is, from outputs B and D. Hence pin 2 goes to pin 9 and pin 3 goes to pin 11.

> (8) To which outputs would you connect the two reset pins on a 7493 counter so that you could count in a scale of (a) three, (b) four, (c) five, (d) six, (e) nine and (f) twelve? (Part (b) might just be tricky!)

The 7490 TTL integrated circuit is a unit which gives a direct divide-by-ten facility, and so is often more convenient than the 7493 which, while it can be made to reset after binary 9, cannot then have its reset facilities available for other purposes – at least, not without the addition of further gates. The pin connections for the 7490 are shown in *Figure 11.19*. The A flip-flop in the 7490 is separate from the other three, as it is in the 7493, so that divide-by-two and divide-by-five functions are available.

By cascading a number of 7490s a counter can be made to count in powers of ten; units, tens, hundreds, etc. A basic circuit is given in *Figure 11.20* which is simply repeated for each decade required. The A input of the first counter is fed with the input pulses; the D output then feeds the A input of the following counter. As the units counter reaches 9 and returns to 0, the D output goes from high to low and the following (tens) counter responds to this transition, recording the 'carry' digit each time. The stages can be reset to zero by momentarily operating switch S which is common to all the connected reset pins and is normally closed. Pins 6 and 7 are not used in this application; they are a pair of reset-to-nine inputs which have a particular application of no interest to us at this point.

By isolating the A flip-flop, division by five is possible with the 7490; the connections this time are indicated in *Figure 11.21*. This circuit, in conjunction with *Figure 11.20*, is useful for a total division

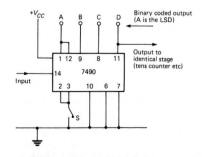

Figure 11.19

Figure 11.20

Sequential logic systems 137

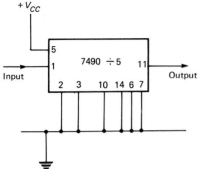

Figure 11.21

by 50, so that if an input at 50 Hz mains frequency (suitably reduced!) is applied at the input, the output from the two dividers will be 1 Hz. This could then drive further counters to record second time intervals. Resetting of this first divider would not normally be used because of the very rapid counting taking place, but if it was required, a normally closed switch would be included in the earth connection of pins 2 and 3.

Numerical Indicators   Calculators, to take an everyday example, would not be very useful if their displays were indicated in binary notation. The insides of the calculators work, of course, in binary, but their outputs need to be in a form that is immediately comprehended: that is, decimal form. We know what 357 means straight away but 101100101 is a different matter!

The usual device available nowadays for the decimal representation of binary numbers (or of words, come to that), is the LED (light emitting diode) or the LCD (liquid crystal display). The LED types are best for general experimental work as they can be bought as single units and arranged as required. The seven-segment variety used for the numerals 0 to 9 is illustrated in *Figure 11.22*. These displays have horizontal and vertical segments lettered *a* to *g* which may be lit up

Figure 11.22

individually to produce the digit required as shown on the right of the figure. In some cases, the upper tail in the 6 and the lower tail in the 9 are omitted. The LEDs that the segments include are brought out to seven connecting pins, with a further pin being the common anode or common cathode connection. If the binary coded output is obtained from a 7490, a decoder is necessary between it and the display. One such decoder suitable for experiments is the 7447A which contains all the logic for switching on the appropriate segments of the display when a binary number is entered.

Figure 11.23 shows the arrangement. The A, B, C and D binary coded inputs from the circuit of *Figure 11.20* are fed in as shown and the display unit is connected via current-limiting resistors to the output. One display unit is required for each 7490/7447 used.

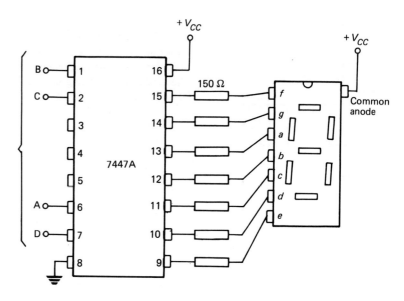

Figure 11.23

CMOS Dividers Dividers are available in the CMOS 4000 range of logic systems and a typical divide-by-ten circuit using the 4017 is shown in *Figure 11.24*.

If a displayed output is required, a separate decoder such as the 7447A is not always necessary as some CMOS counter packages contain an integral decoder and can be connected directly to a LED display. The CMOS 4026 is an example which contains five flip-flops for division purposes and a decoder which provides an output to operate a 7-segment display directly. The 4026 is mounted in a 16-pin DIL package and a suitable circuit is shown in *Figure 11.25*.

The usual handling precautions must be observed when working with CMOS devices. The V_{CC} supply may be anything from 3 V to 15 V; 9 V is usual for circuits driving LED indicators.

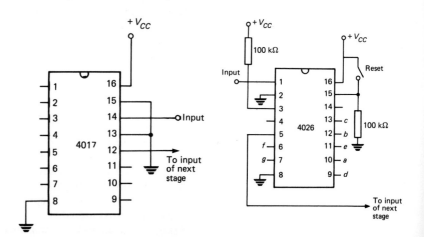

Figure 11.24   Figure 11.25

PROBLEMS FOR SECTION 11

(9) Distinguish between R–S, D-type and J–K flip-flops.

(10) Explain the operation of a flip-flop consisting of two cross-coupled NAND gates.

(11) Fill in the spaces where indicated:
   (a) The basic R–S flip-flop is set by S = 1 and reset by R = ...
   (b) A flip-flop must change rapidly from one .... state to the other by successive ....
   (c) The J–K flip-flop operates on each clock pulse, as determined by the ....
   (d) A group of bits transmitted and received at the same time is called ....

(12) What keeps the outputs of a R–S flip-flop from switching when one input changes to 0 while the other is held at 0?

(13) What is the main difference between a D-latch and a D-flip-flop?

(14) Explain how J–K flip-flops count. How many are needed to count to ten?

(15) Draw a schematic diagram of a binary counter that can count up to decimal 100.

(16) What is the purpose of clock signals in sequential systems?

(17) Two J–K flip-flops which respond to falling edge transitions are connected as shown in *Figure 11.26*. If a 1 kHz square wave is applied at the input, what will be the output?

(18) What is the final output frequency of an 8-stage binary divider with an input of 10 240 Hz?

(19) Draw a circuit for a 7493 package used as a modulo-10 counter.

(20) With the aid of a diagram of each, explain the operation of (a) a 3-bit shift register, (b) a 3-bit counter.

(21) Connect the numbered pins of the integrated circuit shown in *Figure 11.27* to construct an OR gate. All internal gates must be used and the output is to come from pin 8.

(22) The 74164 is described as an 8-bit SIPO shift register. Explain, using diagrams, exactly what this means.

(23) A 7-segment LED display unit was shown in *Figure 11.22* earlier. Decimal 1 is displayed when segments b and c are lit; decimal 2 when segments a, b, g, e, d are lit. Sketch a truth table showing the state of each segment (lit = 1) for the 10 decimal symbols.

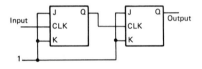

Figure 11.26

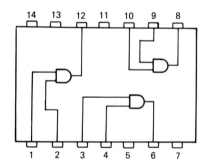

Figure 11.27

Solutions to problems

UNIT SECTION 1

(1) 29

(2) 3

(3) (a) No, a coated wire may be directly heated. (b) No, emission is the same for the same temperatures. (c) No, emission still takes place but electrons return from the cloud and so maintain an equilibrium state.

(4) (a) Neutrons, protons, (b) 1840, (c) valence, (d) hydrogen, (e) cathode material, temperature, (f) holes, (g) conductivity, resistance.

(5) (a) True, (b) true, (c) false.

(6) Your table should look like this:

	Protons	K	L	M
Alum.	13	2	8	3
Silic.	14	2	8	4
Phos.	15	2	8	5
Chlor.	17	2	8	7

(7) There are three valence electrons, therefore gallium impurity is *p*-type.

(8) Germanium; four; four; four.

UNIT SECTION 2

(1) No.

(2) Slightly greater than.

(3) (i) Increases slightly. (ii) Rapidly increases, but is small.

(4) 320 μA.

(5) Yes.

(8) (a) Cathode, (b) space-charge limited (linear), (c) forward, reverse, (d) temperature, (e) same, as.

(9)(a) False, it has only one.

(b) True.

(c) False; all external wires have electron carriers.

(d) True, assuming the temperatures are the same.

(e) True.

(11) About 1.6 mA.

(13) Infinite reverse resistance, zero forward resistance. The curve would simply follow the coordinate axes defining the second quadrant.

(14) Cathode.

(15) No. Thermionic diodes operate at temperatures far above ambient, so changes in ambient have no effect. Also, there are no minority carriers and so no reverse leakage current in the thermionic diode.

(17) (i) 3.0 A in 2 Ω, 1.5 A in 4 Ω; (ii) 1.5 A in 4 Ω, 1.0 A in 6 Ω.

(18) As a capacitor; this application will be covered in the next section.
(19) As a constant voltage source; this is also covered in section 3.
(20) (i) 11.5 mA. (ii) 70.75 V.
(22) As heat at the anode.

UNIT SECTION 3

(1) No, the stated 250 V will be r.m.s. and this has a peak value 354 V.
(2) The peak output is 495 V on each secondary; hence the p.i.v. on the diodes will be twice this, or about 1 kV.
(3) The output voltage will be reduced by the voltage drop across R_f.

$$\text{Average } I = \frac{0.318\,\hat{V}}{R_L + R_f}\,; \qquad \text{r.m.s. } I = \frac{0.5\,\hat{V}}{R_L + R_f}$$

(8) (a) Zero, infinite, (b) zero, (c) space-charge, (d) $\sqrt{(2\,V)}/\pi$, (e) increases, reduces, (f) twice.
(9) 28.3 V.
(10) (a) 0.177 A, (b) 30 W, (c) 0.354 A.
(11) (a) 0.155 A, (b) 31 V, (c) 0.775 A.
(12) (a) 2.44 V, (b) 100 V.
(13) (a) Output voltage would be halved; (b) rectification would cease and the output would be as shown in *Figure A.1*.
(14) (a) 35.4, (b) 1.25 A.
(15) 5.0.
(17) 400 Ω.
(18) 45.5 Ω, 0.35 W.
(19) 44 V maximum to 24.2 V minimum.
(20) 725 Ω.
(21) 24.5 pF.

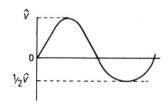

Figure A.1

UNIT SECTION 4

(1) p-n-p.
(2) Common-emitter.
(3) Positive.
(4) 0.1 mA.
(5) (i) 0.96, (ii) 24.
(6) I_E/I_B.
(7) 0.99.
(8) 65.7.
(9) 30–40 Ω; 7000–8000 Ω.
(10) If V_{CB} is reduced to zero, the collector is still able to gather electrons from the base because of the junction p.d. developed across the depletion layer. refer back to page 11.
(11) About 100.
(12) (a) Forward, reverse, (b) electron, (c) negative, (d) away from, (e) common-base, (f) smaller, (g) I_E.
(13) (a) 49, (b) 39, (c) 32.3, (d) 24.
(14) (a) 0.98, (b)).987, (c) 0.991, (d) 0.997.
(15) 0.993, 149.
(16) (a) (i) False, it depends only upon V_{BE}. (ii) False, it depends only upon I_E. (iii) True, for V_{CB} above about 0.2 V.

(b) False, a leakage current flows from collector to emitter.
(c) False; if they were the collector would not collect them.
(d) False.
(17) 6 kΩ.
(18) 200 kΩ.
(19) 1230 Ω.
(20) 13 Ω.
(21) About 1 kΩ.
(23) (a) $\simeq 42$; (b) $\simeq 9$ kΩ

UNIT SECTION 5

(1) A very large V_{CC} would be necessary to allow for the very large voltage drop across R_L. This would appear at the collector when the transistor was OFF and cause damage.
(2) No, input current is in phase with the output current.
(4) Because the equation of the load line is $V_{CC} = V_{CE} + I_C R_L$ and this is the equation of a linear graph.
(5) The ratio of 5.6 mA to 150 µA; about 37 times.
(6) The ratio of 0.15 V input to 8.25 V output; about 55 times.
(7) 0.972.
(8) 0.714 mA.
(9) 180 kΩ.
(10) 2 V.
(13) (a) False, (b) false, V_{BE} decreases, (c) false, it is 10 × 10 = 100, (d) true (strictly for germanium devices), (e) true, (f) false. (When $R_L = 0$ the signal gain is zero.)
(14) (a) 45 µA, (b) 40, (c) 4.8 V peak-to-peak.
(15) $A_v = 1250, A_p = 125\,000$.
(16) (a) 5 kΩ, (b) 6.5 mA.
(17) (a) 5.1 V, (b) 1.2 kΩ.
(18) $A_v = 80, A_i = 33, A_p = 2640$.
(19) $R_L = 500$ Ω, $R_B = 110$ kΩ.
(20) $R_L = 1.2$ kΩ; $R_E = 390$ Ω; $R_1 = 8.6$ kΩ; $R_2 = 2.5$ kΩ.
(21) 11.2 mW, 18.8 mW.

UNIT SECTION 6

(2) Doing the substitution we get $g_{fs} = g_{fso} [1 - V_{GS}/V_p]$. I_{DSS} and V_p are easily found by direct measurement, hence g_{fso} can be found. Then g_{fs} for any V_{GS} can be calculated.
(5) 12.5 kΩ
(7) The gradient at $V_{GS} = 0$ is $2I_{DSS}/V_p$. From *Figure 3.11* it is at once clear that the tangent of the angle of gradient is $I_{DSS}/\frac{1}{2}V_p$ which gives the required proof.
(8) $r_d = 6.25$ kΩ; $g_{fs} = 6$ mS, $\mu = 37.5$.
(13) 3 mA, assuming V_{DS} remains constant.
(14) 66.7 kΩ.
(15) $\mu = 44$; $g_{fs} = 3.3$ mS, $r_d = 13.3$ k.
(16) (a) thermionic valves; (b) depletion; (c) n- or p-channel construction; (d) zero; (e) source to drain.
(17) $\mu = 10$; $A_v = 16.7$
(18) About 15 kΩ
(19) (a) 0.5 V; (b) 1.48 mS; (c) 470 Ω; (d) 6750 Ω.
(20) $r_d \simeq 6.67$ kΩ; $g_{fs} \simeq 1.7$ mS
(21) $R_1 = 3$ MΩ; $R_2 = 1$ MΩ. $A_v \simeq 7.5$.
(22) About 2.0 mS.

UNIT SECTION 7

(1) Transistor T_2 would turn off.
(5) 267 Ω.
(6) 46 Ω; 0.35 W.
(7) 62.5 mW; 10.3 mA.
(8) 180 mV.

UNIT SECTION 8

(3) $A+B+C$ is represented by three parallel connected switches. The required truth table follows:

A	B	C	F
0	0	0	0
0	0	1	1
0	1	0	1
0	1	1	1
1	0	0	1
1	0	1	1
1	1	0	1
1	1	1	1

(4) $A+(B.C) = F$.
(5) $2^4 = 16$; $A(B+C)+D = F$
(8) The circuits are shown in *Figure A.2*.

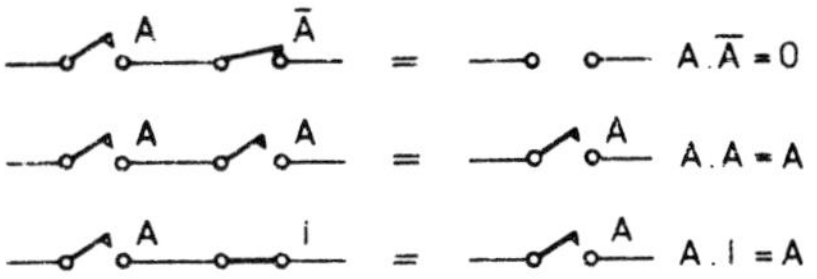

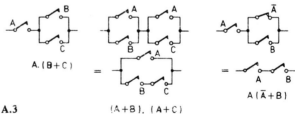

Figure A.2

(11) $A+B = F$; $A.B = F$; $A.B + \bar{A}.\bar{B} = F$.
(12) The circuits are shown in *Figure A.3*.

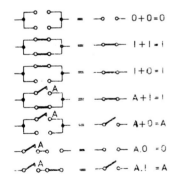

Figure A.3 $(A+B).(A+C)$

(13) AND.
(14) $A[B+C(D+\bar{E})]$.
(16) $A.B(C+A).(B+A.C+D).E$.
(17) $B.\bar{C}+\bar{B}.C$.
(18) The circuits are shown in *Figure A.4*.

Figure A.4

UNIT SECTION 9

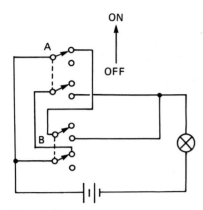

Figure A.5

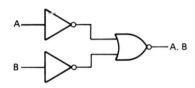

Figure A.6

(1) (a) Coincidence, (b) series, (c) logical 0
(2) See *Table 8.2* on page 85.
(3) *Figure A.5* shows one possible solution.
(4) The truth table is as follows:

A	B	C	F
0	0	0	0
0	0	1	1
0	1	0	1
0	1	1	0
1	0	0	1
1	0	1	0
1	1	0	0
1	1	1	0

(5) (a) 4-5 ms, 8-9 ms; (b) at all times except 7-8 ms; (c) 23 ms, 6-7 ms, 9-10 ms.
(6) Forward
(11) See *Figure A.6*.
(12) See *Figure A.7*.
(13) (c) and (d)
(14) (b)
(17) (a) Logical 1; (b) EX-OR; (c) OR, NOT; (d) NOR, NAND; (e) AND
(19) (a) $\bar{A}$; (b) A.B; (c) A+B; (d) A $\oplus$ B; (e) $AB+\bar{A}\bar{B}$
(22) The first circuit will not work because the output will follow the B input irrespective of conditions at A.
(23) (a) When A or B are 10 V high; (b) about zero; (c) NOR
(24) See *Figure A.8*.
(29) The AND gate is faulty.
(30) Red on when $A(\bar{B}+C)$; green on when $ABC\bar{D}$.

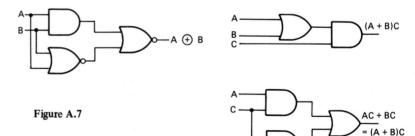

Figure A.7

Figure A.8

Solutions to problems 145

UNIT SECTION 10

(1) 8 cells. $F = \overline{A}\overline{B}\overline{C} + \overline{A}B\overline{C} + A\overline{B}\overline{C} + ABC + \overline{A}BC$
$= \overline{B}\overline{C} + BC + \overline{A}B\overline{C}$
(3) $F = A(\overline{B} + \overline{C})$
(4) (c) only
(10) (a) $A.B$ (b) $AC + \overline{A}B$ (c) $AB + BC$ (d) $A + B$
(e) $\overline{A}C + BC$ (f) $B + \overline{A}\overline{C}$ (g) $AD + A\overline{B} + \overline{A}C$
(11) $\overline{B}.\overline{D}$
(12) $R = C, S = \overline{A}\overline{B}CD, P = B\overline{C}, Q = A\overline{B}\overline{C}$
(13) See *Figure 10.14*.
(14) $\overline{A}\overline{B} + ABC$

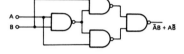

Figure A.9

(15) See *Figure A.9*.
(16) $ABD + ACD + BCD$
(17) $A + B$. Use two NOR gates in cascade.

UNIT SECTION 11

(1) Short collector of OFF transistor to earth.
(3) See Table:

In	Q	$\overline{Q}$
0	0	1
0	1	0
1	1	0
1	0	1
0	0	1

(6) $2^9 = 512$, therefore use 9 stages.
(7) (a) Q outputs on B and C; $\overline{Q}$ outputs on A and D.
  (b) Q outputs on A, B and D; $\overline{Q}$ output on C.
(8) (a) A, B (b) V_{CC}, C (c) A, C (d) B, C
  (e) A, D (f) C, D.
(11) (a) 0 (b) Stable; clock pulses
  (c) Inputs (d) parallel data.
(14) 4
(16) For purposes of synchronising different parts of a system.
(17) 250 Hz square wave
(18) 40 Hz
(21) Connect 1 to 2 as input A; connect 3 to 4 as input B; connect 6 to 9; connect 10 to 12; take output from 8.
(23) See table:

Decimal	a	b	c	d	e	f	g
0	1	1	1	1	1	1	0
1	0	1	1	0	0	0	0
2	1	1	0	1	1	0	1
3	1	1	1	1	0	0	1
4	0	1	1	0	0	1	1
5	1	0	1	1	0	1	1
6	1	0	1	1	1	1	1
7	1	1	1	0	0	0	0
8	1	1	1	1	1	1	1
9	1	1	1	1	0	1	1

Appendix

GRADIENTS

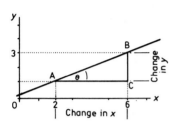

Figure A.10

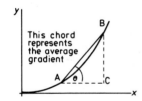

Figure A.11(a)

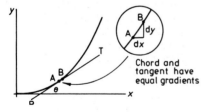

Figure A.11(b)

It is frequently necessary to be able to measure or calculate the *gradient* or the *slope* of a graph which represents the relationship between two variable quantities. We define the gradient of a line (see *Figure A.10*) as the tangent of the angle (θ) which the line makes with the horizontal axis. This can be found quite easily by constructing a triangle ABC as shown in the figure and applying the tangent ratio:

$$\tan \theta = \frac{BC}{AC} = \frac{\text{Change in } y}{\text{Change in } x}$$

Hence, if x changed from 2 to 6, and y correspondingly changed from 1 to 3 as the figure shows, then

$$\tan \theta = \frac{3-1}{6-2} = 0.5$$

from which, by the use of tables, $\theta = 26.6°$. However, we are not so much interested in the actual angle as in the ratio given by 0.5; this tells us the relationship existing between changes in x and the corresponding changes in y. In this example, we know that y is changing at only half the rate of x.

Obviously, if the graph is a straight line, the value obtained for the gradient would be unaffected by the size or position of the triangle ABC. The gradient of the line, in other words, is constant, a fairly self-evident fact. What happens in the case of a graph which is not a straight line, but is curved as shown in *Figure A.11(a)*? Here the gradient is not constant but changes all the time. In this case, we must define the gradient at any given point. If we choose two points on the curve, as at A and B, and complete the triangle ABC, the average gradient between A and B is given by the tangent of angle θ. Imagine now that B moves down the curve towards A; the chord AB reduces in length, and the corresponding changes in x and y also reduce. When B is extremely close to A, as in diagram (*b*), we can consider the chord AB to be coincident with the tangent PT drawn to the curve at the point in question. Hence the *gradient of the curve at this point* is given as the *gradient of the tangent PT*. As the changes in x and y are now also very small, we refer to them as δx and δy respectively, meaning 'the small change in x' and 'the small change in y' respectively. Hence

$$\text{Gradient at a point} = \frac{\text{Small change in } y}{\text{Small change in } x} = \frac{\delta y}{\delta x}$$

and this ratio is identical with the gradient of the tangent drawn to the curve at any point. It is, in fact, the *instantaneous* gradient of the curve, or the instantaneous rate of change of the variable quantities.

You will learn mathematical processes for finding the gradient of a curve at any given point when you meet up with elementary differential calculus in your mathematics course. For the present, the process of drawing a tangent to the curve at the point in question will enable you to work out what the gradient is at that point by simple triangle methods as we have just discussed.

Index

Acceptor impurities, 8
Alloy-diffused diode, 11
Alloy-diffused transistor, 32
Alpha current gain, 34
Amplifier:
    common-emitter, 44
    FET, 66 et seq.
    power, 44
    small-signal, 44 et seq.
Amplification factor, of FET, 67
AND gate, 98
Anode, thermionic valve, 14
Antimony, 7
Arsenic, 7
Asynchronous counter, 133
Atoms:
    pentavalent, 7
    tetravalent, 2
    trivalent, 8
Avalanche effect, 13

Barrier potential, 11
Base bias, 45
Base, of transistor, 31
Bipolar transistor, 31
Bistable flip-flop:
    clocked, 127
    D-type, 128
    J–K, 128
    integrated, 126
Boolean algebra:
    relationships in, 91
    in switching systems, 90
    theorems of, 92
'Bottoming', 55
Breakdown voltage, 13
Bridge rectifier, 21

Capacitance diode, 29
Carriers:
    majority, 7
    minority, 10
    thermally generated, 6
Cathode, 4
Channel, of FET, 60
Charge:
    on electron, 2
    on proton, 2
Class-A amplifier, 46
Clock pulse, 128
Clocked flip-flop, 128
CMOS technology, 110
Coincidence gate, 98
Common-base mode, 33

Common-collector mode, 33
Common-emitter mode, 33
Common-source amplifier, 68
Conductance, mutual, 55
Conduction:
    extrinsic, 10
    intrinsic, 7
Conductivity:
    n-type, 7
    p-type, 8
Counter, bistable:
    scale of 10, 136
    scale of 12, 135
    scale of 16, 135
Counting circuits, 136
Crystalline structure, 5
Current gain, 34

D-latch, 128
De Morgan's Rules, 95
Depletion layer, 11
Diffused junction, 11, 32
Diode:
    characteristics, 13, 15
    p-n junction, 11
    semiconductor, 11
    steering, 125
    as switch, 97
    symbol, 12
    thermionic, 14
Diode logic, 98 et seq.
Diode-resistor logic, (DRL), 98, 99
Diode-transistor logic, (DTL), 103
Divider, bistable as, 123
Divider, CMOS, 138
Donor impurities, 7
'Doping', 7
Drain, of FET, 57
Dynamic load line, 48
Dynamic operating point, 46

Electron:
    hole pair, 7
    particle, 1
    thermally generated, 6
Emitter, 31
Emission, thermionic, 3
Enhancement FET, 73
Equivalent circuit:
    of bipolar transistor, 55
    of FET, 69
Exclusive-OR gate, 96, 99
Extrinsic conductivity, 10

Field-effect transistor, 60 *et seq.*
Filament, 5
Flip-flops:
   clocked, 127
   D-latch, 128
   integrated, 126
   J–K, 128
   R–S, 124
   triggering, 125
Free electrons, 3
Full-wave rectifier, 20

Gain:
   current, 34
   power, 57
   voltage, 57
Gate circuits, 96 *et seq.*
Gate, of FET, 56
Germanium atom, 2
Gradient of curve, 146
Grounded emitter, 33

Half-wave rectifier, 20
Handling CMOS, 111
Harmonic distortion, 44
Heat sinks, 57
Helium atom, 2
Hole, 6
Hydrogen atom, 2

Ideal diode, 21
Impurities:
   n-type, 7
   p-type, 7
Indicators, numerical, 137
Indium, 32
Input characteristic:
   FET, 67
   transistor, 36
Insulated-gate FET (IGFET), 72
Integrated logic, 105 *et seq.*
Intrinsic conductivity, 7
Inverting amplifier, 55
Ion, 3
Ionisation, 3

J–K bistable flip-flop, 128
Junction-gate transistor (JUGFET), 60
Junction potential barrier, 11

Karnaugh mapping, 114 *et seq.*
   application, 120
   minimisation, 118
   structure, 115

Lattice structure, 5
Leakage current, 12, 51
Light-emitting diode (LED), 137
Liquid crystal display (LCD), 137

Load line:
   bipolar transistor, 49
   FET, 71
Logic gates:
   AND, 98
   coincidence, 98
   EXCLUSIVE-OR, 96, 99
   Integrated, 106
   NAND, 101, 103, 104
   NOR, 101, 103
   NOT, 97
   OR, 99
Logical algebra, 90
Logical switching, 85 *et seq.*

Majority carriers, 7
Mass:
   electron, 2
   proton, 2
Minority carriers, 10
MOSFET:
   depletion, 67
   enhancement, 67
   symbol, 73
Multiple emitter, 103
Multivibrator, bistable, 126
Mutual conductance, 55, 64

n-type semiconductor, 7
NAND gate, 101, 103, 104
Negation, 87
Negative logic, 84
Neutrons, 2
Noise margin, 108
NOR gate, 101, 103
NOT gate, 97
Nucleus, 2

Open-collector output, 109
Operating point, 46
OR gate, 99
Output characteristic:
   FET, 63
   transistor, 38
Oxide coating of cathode, 5, 14

p-n diode, 11
p-type semiconductor, 7
Peak inverse voltage, 22
Pentavalent atoms, 7
Phosphorus, 7
Pinch-off voltage, 62
PISO registers, 132
Positive logic, 84
Potential barrier, 11
Proton:
   mass, 2
   charge, 2

Quartz crystal, 6

R–S flip-flop, 124
Recombination, 33
Rectifiers:
   bridge, 21
   full-wave, 20
   half-wave, 19
Reservoir capacitor, 23
Resistor-transistor logic (RTL), 103
Reverse bias, 12
Ripple voltage, 23
Rules of logical algebra, 92

S–R flip-flop, 124
Saturation, 39
Seven-segment display, 137
Shell, electron, 2
Shift register, 130
Silicon atom, 2
Sink current, 108
SIPO register, 132
Slope resistance, 40
Smoothing, 23
Source current, 108
Space charge, 4, 14
Stability of operating point, 52
Stabilized power supplies:
   integrated, 80
   series controlled, 79
   shunt controlled, 78
   zener diode, 25 *et seq.*
Static characteristics:
   diode, 13
   FET, 63 *et seq.*
   transistor, 47 *et seq.*
Static load line, 49, 71
Substrate, 73
Switching:
   diode, 19, 97
   transistor, 55

Tetravalent atoms, 8
Thermal conduction, 6
Thermal runaway, 53, 56
Thermionic diode, 14
Thermionic emission, 3
Totem pole output, 104
Transfer characteristics, 40, 64
Truth tables, 84 *et seq.*
TTL, 103 *et seq.*
Tungsten, 4

Unidirectional current, 19
Unipolar transistor (FET), 60 *et seq.*

Valence bonds, 6
Valence electrons, 3
Varactor diode, 29
Voltage comparator, 79
Voltage gain, 57, 68
Voltage regulator diode, 25

Zener breakdown, 25
Zener diode, 24
Zener protection, 28